Analog Circuits and Signal Processing

Series editors

Mohammed Ismail, Dublin, USA
Mohamad Sawan, Montreal, Canada

For further volumes:
http://www.springer.com/series/7381

Francesco Brandonisio
Michael Peter Kennedy

Noise-Shaping All-Digital Phase-Locked Loops

Modeling, Simulation, Analysis and Design

Springer

Francesco Brandonisio
Tyndall National Institute
Cork
Ireland

Michael Peter Kennedy
University College Cork
Cork
Ireland

ISSN 1872-082X ISSN 2197-1854 (electronic)
ISBN 978-3-319-34441-6 ISBN 978-3-319-03659-5 (eBook)
DOI 10.1007/978-3-319-03659-5
Springer Cham Heidelberg New York Dordrecht London

Printed on acid-free paper

Springer is part of Springer Science+Business Media (www.springer.com)

To
Rocco, Nuccia
and
Rossana

Preface

All-Digital Phase Locked Loops (ADPLLs) have become very common in low cost and feature mobile phones. In recent years, extensive research activity on ADPLLs has focused on increasing the performance of ADPLLs, thus increasing the range of their possible applications. The theoretical noise performance of an ADPLL is limited by the quantization error of the TDC and the DCO. In the literature, noise shaping of the quantization error is one of the main techniques used to reduce the impact of the quantization error of the TDC and the DCO on the noise performance of an ADPLL. In this book, we present a framework to analyse, design, simulate and compare different ADPLL architectures with noise shaping TDCs and DCOs.

In Chap. 1, we summarize the main contributions of the book.

In Chap. 2, we review the operations of the main ADPLL architectures in terms of phase-to-digital conversion.

In Chap. 3, we review the main TDC architectures and relate their operations to quantizers and/or sigma-delta modulators.

In Chap. 4, we derive discrete-time models for the main ADPLL architectures and derive analytical equations for predicting the phase noise performance.

In Chap. 5, we show the advantages of noise shaping and dither by means of an analytical method in the time domain.

In Chap. 6, we focus on simulating ADPLLs as mixed-signal systems. We show that there is a tradeoff between accuracy and simulation time. We describe a simulation method in Simulink that can mitigate this tradeoff.

In Chap. 7, we discuss phase noise in more detail. We describe a procedure to model and extract the phase noise of a signal in Matlab.

This work was supported in part by Science Foundation Ireland under grant 08/IN.1/I854 and by Microelectronic Circuits Centre Ireland.

Cork, September 2013

Francesco Brandonisio
Michael Peter Kennedy

Contents

Acronyms

ADC	Analog-to-Digital Converter
ADPLL	All-Digital Phase Locked Loop
DC	Direct Current; here it denotes a constant signal
DCO	Digitally Controlled Oscillator
DNL	Differential Nonlinearity
EMF	Embedded Matlab Function
FPGAy	Field Programmable Gate Array
GRO	Gated Ring Oscillator
INL	Integral Nonlinearity
LTI	Linear Time Invariant
PFD	Phase Frequency Detector
PLL	Phase Locked Loop
PSD	Power Spectral Density
TDC	Time-to-Digital Converter
VCO	Voltage-Controlled Oscillator

Chapter 1
Introduction

An All Digital Phase Locked Loop (ADPLL) is an alternative to a traditional Phase Locked Loop (PLL) for implementation in nanoscale digital CMOS, especially as part of a system-on-chip (SoC) [1, 2]. One of the key advantages of ADPLLs over their analog counterparts is that they remove the need for large capacitors within the loop filter by utilizing digital circuits to achieve the desired filtering function. The resulting area savings are important for achieving a low-cost solution. Moreover, the phase error signal in an ADPLL is a digital word that is produced by means of digital circuits such as accumulators, samplers and Time-to-Digital Converters [3, 4]. As a consequence, a much more attractive, mostly digital, design flow is achieved [2]. Furthermore, the essentially digital architecture of an ADPLL can be augmented with reconfigurable gains and a filter [5, 6], or with a software-assisted digital processor for calibration [7].

In the literature of recent years, there has been a significant effort to improve the performance of ADPLLs in terms of the output frequency [6, 8], resolution [9], bandwidth [10–12], locking speed [13, 14], phase noise [15–17], and power consumption [18].

In this book, we focus on ADPLLs that include TDCs with noise-shaping of the quantization error. We explain how to design an ADPLL, analyse its noise performance and realize behavioral models that require the minimum possible simulation time. We also review the main TDC architectures in order to highlight the advantages of a first-order noise-shaping TDC. We show how to relate the operation of a TDC to a quantizer and/or a first-order sigma-delta modulator. Furthermore, we determine analytically the precisions of a quantizer and a sigma-delta modulator when followed by a moving average filter in terms of the maximum difference between the input and the output when this difference is bounded.

First-order noise-shaping of the quantization error is produced by a system that can be represented by a first-order sigma-delta modulator. During our analysis, we show that there is an *inherent* sigma-delta modulation in both Vernier TDCs and accumulator-based ADPLLs. In fact, we demonstrate that the models of a Vernier TDC and an accumulator-based ADPLL can be based on sigma-delta modulators. In

F. Brandonisio and M. P. Kennedy, *Noise-Shaping All-Digital Phase-Locked Loops*, Analog Circuits and Signal Processing, DOI: 10.1007/978-3-319-03659-5_1,

particular, we use the equations of a sigma-delta modulator to implement an efficient model of an accumulator-based ADPLL.

The book is organized as follows:

In Chap. 2, we review the operating principles of the main topologies of ADPLLs (PFD-plus-TDC-based, TDC-based and accumulator-based ADPLLs) in terms of the integer and fractional parts of the phase difference. We also mention the flip-flop-based ADPLL which can be considered as a special case of a TDC-based ADPLL. We show models that describe the phase-to-digital conversion in each ADPLL architecture. We also discuss possible strategies to clock the digital filter in the various ADPLL architectures. We show how to modify the ADPLL architectures to synthesize a fractional ratio between the frequencies of the reference oscillator and the DCO. Finally, we compare the ADPLL architectures in terms of phase-to-digital conversion, TDC dynamic range, and metastability. The analytical approach used in this chapter is general and it can be extended to any modified ADPLL architecture that can be obtained from those we have studied.

In Chap. 3, we briefly review the main TDC architectures that have been published in the literature. We introduce notation for comparing different TDC architectures in terms of their operating principles and time resolution. By comparing different TDC architectures, we show that a first-order noise-shaping TDC is an interesting architecture for digital systems because it allows a tradeoff between high time resolution and speed. We also show how the Vernier method can be considered as a special case of sigma-delta modulation. To our knowledge, the Vernier method is recognized as a particular case of sigma-delta modulation for the first time in this work. A Vernier-TDC model based on a sigma-delta modulator is also new in the literature, to our knowledge. There are already reviews of TDC architectures such as [19] and [20]. However, in this review, a *single* notation is used to derive models based on quantizers and sigma-delta modulators for the main TDC architectures from their respective timing diagrams. The same notation allows a simple comparison between different architectures in terms of time resolution.

In Chap. 4, we focus on the noise performance of the main architectures of ADPLLs. We derive analytical predictions of the phase noise in TDC-based and accumulator-basedADPLLs with noise-shaping TDCs and a DCO driven by a sigma-delta modulator. In order to derive analytical predictions for the ADPLL phase noise, we first explain how to calculate the phase noise of the DCO when its input is known. Then we derive linear models associated with the building blocks in an ADPLL. We use the linear models of the building blocks of an ADPLL to develop a linear model of the full ADPLL. We derive analytical predictions of the ADPLL phase noise from the linear ADPLL model. Finally, we compare Matlab simulations and analytical predictions for an example TDC-based ADPLL architecture. We also show that our results are in very good agreement with predictions obtained by means of the "PLL Design Assistant" [21] which is an automatic design tool for PLL. By contrast with the "PLL Design Assistant" program, which is compiled and therefore cannot be modified by the user, the Matlab scripts that we present are ready to be edited. Hence, the Matlab scripts in this chapter represent a complementary learning tool that gives direct insight into the design equations.

In Chap. 5, we determine the precision of the systems "Quantizer plus Moving Average Filter" and "Sigma-Delta Modulator plus Moving Average Filter" with dither. We show analytically that the difference between the input and output of a "Sigma-Delta Modulator plus Moving Average Filter" is smaller than that of an equivalent "Quantizer plus Moving Average Filter". The analytical results derived in this chapter are important in order to understand how to exploit noise shaping in ADPLLs. Notice that we determine the precision of systems comprising sigma-delta modulation and dither followed by moving average filters in terms of the maximum difference between the input and the output and the variance of the output in the *time domain*. In the literature, it is common to analyse the advantages of sigma-delta modulation and dither in the *frequency domain* [22]. However, the analysis in the frequency domain is usually based on the assumptions that the quantization error of a sigma-delta modulator is white and independent of the input. This white noise approximation is not valid when the output of a sigma-delta modulator exhibits tones. The approach that is presented in this chapter is an alternative to the standard frequency domain analysis and does not require the white noise approximation.

In Chap. 6, we discuss how to simulate an ADPLL with fully nonlinear behavioral models. We show how to realize an efficient behavioral model of an ADPLL that produces the *minimum* number of samples during a simulation. The equations that we use to implement the efficient model of an ADPLL are related to a sigma-delta modulator. We report example C- and Matlab code that can be used to implement an efficient Simulink model of an ADPLL. We also illustrate how to realize a Simulink S-function that controls the simulation loop while the simulation is running. The modelling approach for ADPLLs that is detailed in this chapter is similar to that described by Staszewski et al. [23]. The technique reported in [23] describes how to realize event-driven models in Verilog-AMS. Models that are event-driven can be built in Verilog and Verilog-AMS by means of the command "timer". In this chapter, we show how to realize event-driven models in Simulink. The Simulink models that we describe are simple and can be used as learning tools to understand how to deal with the problems that are associated with modeling ADPLLs.

In Chap. 7, we show how to model and calculate the phase noise of an oscillator in Matlab. The approach presented has been adopted in Chap. 4 to calculate the phase noise of an ADPLL. The material included in this chapter refers mostly to Kundert's work on model oscillators and signals with phase noise in Verilog-AMS [24]. However, we focus more on the steps that are necessary to model a noisy signal and to extract its phase noise. The steps of an example noise extraction procedure are clearly illustrated with Matlab scripts.

References

1. M. Kumm, H. Klingbeil, P. Zipf, An FPGA-Based Linear All-Digital Phase-Locked Loop, IEEE Trans. Circuits Sys. I: Regular Papers, vol. 57, no. 9, pp. 2487–2497, Sep. 2010.
2. R. B. Staszewski, State-of-the-Art and Future Directions of High-Performance All-Digital Frequency Synthesis in Nanometer CMOS, IEEE Trans. Circuits Sys. I: Regular Papers, vol. 58, no. 7, pp. 1497–1510, Jul. 2011.

3. R. B. Staszewski, P. T. Balsara, All-Digital Frequency Synthesizer in Deep-Submicron CMOS, John Wiley & Sons Inc., 2006.
4. R. B. Staszewski, P, T. Balsara, Phase-Domain All-Digital Phase-Locked Loop, IEEE Trans. Circuits Sys. II: Express, Briefs, vol. 52, no. 3, Mar. 2005.
5. D.-S. Kim, H. Song, T. Kim, S. Kim, D.-K. Jeong, A 0.3–1.4 GHz All-Digital Fractional-N PLL With Adaptive Loop Gain Controller, IEEE J. Solid-State Circuits, vol. 45, no. 11, pp. 2300–2311, Nov. 2010.
6. S.-Y. Yang, W.-Z. Chen, T.-Y. Lu, A 7.1 mW, 10 GHz All Digital Frequency Synthesizer with Dynamically Reconfigured Digital Loop Filter in 90 nm CMOS Technology, IEEE J. Solid-State Circuits, vol. 45, no. 3, pp. 578–586, Mar. 2010.
7. R. Staszewski, R. B. Staszewski, T. Jung, T. Murphy, I. Bashir, O. Eliezer, K. Muhammad, M. Entezari Software Assisted Digital RF Processor (DRP^{TM}) for Single-Chip GSM Radio in 90 nm CMOS, IEEE J. Solid-State Circ., vol. 45, no. 2, pp. 276–288, Feb. 2010.
8. C.-C. Hung, S.-I. Liu, A 40-GHz Fast-Locked All-Digital Phase-Locked Loop Using a Modified Bang-Bang Algorithm, IEEE Trans. Circuits Sys. II: Express Briefs, vol. 58, no. 6, pp. 321–325, Jun. 2011.
9. S.-K. Lee, Y.-H. Seo, H.-J. Park, J.-Y. Sim, A 1 GHz ADPLL with a 1.25 ps Minimum-Resolution Sub-Exponent TDC in 0.18 μm CMOS, IEEE J. Solid-State Circuits, vol. 45, no. 12, pp. 2874–2881, Dec. 2010.
10. M. Zanuso, S. Levantino, C. Samori, A. L. Lacaita, A Wideband 3.6 GHz Digital $\Delta\Sigma$ Fractional-N PLL With Phase Interpolation Divider and Digital Spur Cancellation, IEEE J. Solid-State Circuits, vol. 46, no. 3, pp. 627–638, Mar. 2011.
11. E. Temporiti, C. Weltin-Wu, D. Baldi, M. Cusmai, F. Svelto, A 3.5 GHz Wideband ADPLL With Fractional Spur Suppression Through TDC Dithering and Feedforward Compensation, IEEE J. Solid-State Circuits, vol. 45, no. 12, pp. 2723–2736, Dec. 2010.
12. E. Temporiti, C. Weltin-Wu, D. Baldi, R. Tonietto, F. Svelto, A 3 GHz Fractional All-Digital PLL With a 1.8 MHz Bandwidth Implementing Spur Reduction Techniques, IEEE J. Solid-State Circuits, vol. 44, no. 3, pp. 824–834, Mar. 2009.
13. X. Chen, J. Yang, L.-X. Shi, A Fast Locking All-Digital Phase-Locked Loop via Feed-Forward Compensation Technique IEEE Trans. Very Large Scale Int. (VLSI) Sys., vol. 19, no. 5, pp. 857–868, May. 2011.
14. C.-T. Wu, W.-C. Shen, W. Wang, A.-Y. Wu, A Two-Cycle Lock-In Time ADPLL Design Based on a Frequency Estimation Algorithm, IEEE Trans. Circuits Sys. II: Express, Briefs, vol. 57, no. 6, Jun. 2010.
15. H.-J. Hsu, S.-Y. Huang, A Low-Jitter ADPLL via a Suppressive Digital Filter and an Interpolation-Based Locking Scheme IEEE Trans. Very Large Scale Int. (VLSI) Sys., vol. 19, no. 1, pp. 165–170, Jan. 2011.
16. T. Tokairin, M. Okada, M. Kitsunezuka, T. Maeda, M. Fukaishi, A 2.1–2.8-GHz Low-Phase-Noise All-Digital Frequency Synthesizer With a Time-Windowed Time-to-Digital Converter, IEEE J. Solid-State Circuits, vol. 45, no. 12, pp. 2582–2590, Dec. 2010.
17. C.-M. Hsu, M. Z. Straayer, M. H. Perrott, A Low-Noise Wide-BW 3.6-GHz Digital $\Delta\Sigma$ Fractional-N Frequency Synthesizer With a Noise-Shaping Time-to-Digital Converter and Quantization Noise Cancellation, IEEE J. Solid-State Circuits, vol. 43, no. 12, pp. 2776–2786, Dec. 2008.
18. L. Xu, S. Lindfors, K. Stadius, Jussi RyyRnen, A 2.4-GHz Low-Power All-Digital Phase-Locked Loop, IEEE J. Solid-State Circuits, vol. 45, no. 8, pp. 1513–1521, Aug. 2010.
19. G.W. Roberts, M. Ali-Bakhshian, A Brief Introduction to Time-to-Digital and Digital-to-Time Converters, IEEE Trans. Circuits Sys. II: Express, Briefs, vol. 57, no. 3, Mar. 2010.
20. J. Kalisz, Review of Methods for Time Interval Measurements with Picosecond Resolution, Metrologia, vol. 41, no. 1, pp. 1732, Feb. 2004.
21. PLL Design Assistant Program, a design tool developed by Mike Perrott and available at http://www.cppsim.com Accessed in Nov. 2012

22. R. Schreier, G. C. Temes, Understanding Delta-Sigma Data Converters, IEEE Press, John Wiley & Sons Inc., 2005.
23. R.B. Staszewski, P.T. Balsara, C. Fernando, Event-driven Simulation and modeling of phase noise of an RF oscillator, IEEE Trans. Circuits Sys. I: Regular Papers, vol. 52, no. 4, Apr. 2005.
24. K. Kundert, Predicting the Phase Noise and Jitter of PLL-Based Frequency Synthesizers, Designer's Guide Consulting Inc., downloadable at http://www.designers-guide.org/analysis/PLLnoise+jitter.pdf. Last access in October 2012

Chapter 2
Phase Digitization in All-Digital PLLs

2.1 Introduction

In this chapter, we review the operating principles of the main topologies of ADPLLs (PFD-plus-TDC-based, TDC-based and accumulator-based ADPLLs) in terms of the integer and fractional parts of the phase difference. We also mention the flip-flop-based ADPLL which can be considered as a particular case of a TDC-based ADPLL. We show models that describe the phase-to-digital conversion in each ADPLL architecture when the integer part of the phase difference is equal to or different from zero. We show that a flip-flop based ADPLL can be viewed as the ADPLL architecture with the simplest phase-to-digital conversion. We also discuss possible strategies to clock the digital filter in the various ADPLL architectures. We show how to modify the ADPLL architectures to synthesize a fractional ratio between the frequencies of the reference oscillator and the DCO. Finally, we compare the ADPLL architectures in terms of phase-to-digital conversion, TDC dynamic range, and metastability.

2.2 Definitions of Integer and Fractional Phases

The phase $\boldsymbol{\Phi}(t)$ of a signal at the instant t is the integral of the instantaneous frequency over the time interval $[0, t]$, as shown by the following equation:

$$\boldsymbol{\Phi}(t) = \int_{\tau=0}^{t} f(\tau)d\tau + \boldsymbol{\Phi}_0, \tag{2.1}$$

where $f(t)$ and $\boldsymbol{\Phi}_0$ are the instantaneous frequency and the initial phase of the signal. According to Eq. (2.1), the phase of a signal is determined when the initial phase and the frequency $f(t)$ over the interval $[0, t]$ are known. In order to simplify our analysis, we assume that the signal with unknown frequency can be associated with

F. Brandonisio and M. P. Kennedy, *Noise-Shaping All-Digital Phase-Locked Loops*, Analog Circuits and Signal Processing, DOI: 10.1007/978-3-319-03659-5_2,

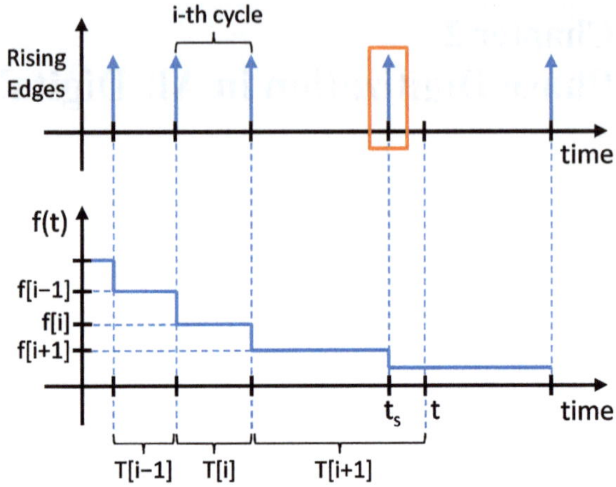

Fig. 2.1 Timing diagram showing the cycles and the instantaneous frequency of a signal

cycles. The cycles are determined by choosing a reference level which is crossed by the signal. The beginning and the end of a cycle are determined by two consecutive crossings. Without loss of generality, we assume that the rising edges of a signal are associated with the beginnings and the ends of the cycles. Moreover, we assume that the frequency of the signal is *constant* and equal to $f[i]$ over the i-th cycle. Hence, the duration of the i-th cycle is equal to the period $T[i] = 1/f[i]$ of the signal. The frequency $f[i]$ can be obtained by measuring the duration $T[i]$ of the i-th cycle. Without loss of generality, we assume that the rising edges of a signal determine the durations of the cycles that have to be measured. Figure 2.1 shows the cycles and the values of the instantaneous frequency of an example signal.

When the frequency $f(t)$ is a piecewise constant function of the type shown in Fig. 2.1, Eq. (2.1) can be simplified, as we will show in the following. We first write Eq. (2.1) as:

$$\boldsymbol{\Phi}(t) = \int_{\tau=0}^{t_s} f(\tau)d\tau + \int_{\tau=t_s}^{t} f(\tau)d\tau, \tag{2.2}$$

where t_s is the instant associated with the previous rising edge with respect to t, as shown in Fig. 2.1. There are two integrals in Eq. (2.2). By considering that $f(t)$ is equal to $1/T[k]$ between two consecutive rising edges of the k-th cycle, the first integral in Eq. (2.2) gives:

$$\int_{\tau=0}^{t_s} f(\tau)d\tau = \sum_{i=1}^{n} f[i] \cdot T[i] = n, \tag{2.3}$$

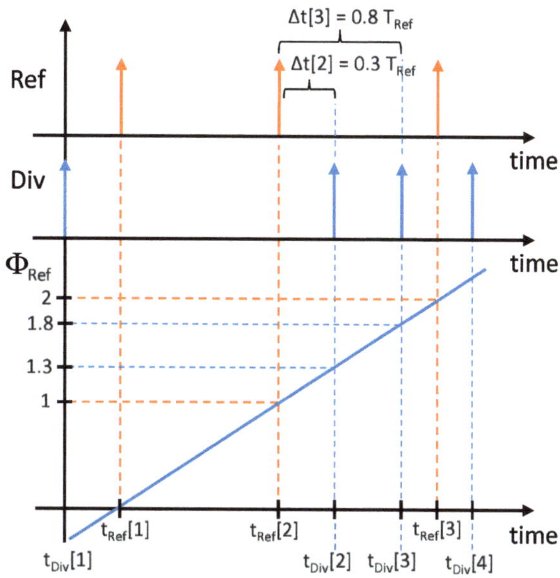

Fig. 2.2 Phase of the signal *Ref* expressed in cycles

where n is the number of *complete* cycles of the signal over the time interval $[0, t]$. We define the integer part of the phase $\boldsymbol{\Phi}_{int}(t)$ of a signal as the number of *complete* cycles of the signal over the time interval $[0, t_s]$.

By considering that $f(t)$ is equal to $1/T[n+1]$ over the time interval $[t_s, t]$, the second integral in Eq. (2.2) gives:

$$\int_{\tau=t_s}^{t} f(\tau)d\tau = \frac{(t-t_s)}{T[n+1]}. \tag{2.4}$$

We define the fractional part of the phase $\boldsymbol{\Phi}_{frac}(t)$ as the ratio $(t-t_s)/T[n+1]$. We conclude that the phase $\boldsymbol{\Phi}(t)$ of a signal can be written as:

$$\boldsymbol{\Phi}(t) = \boldsymbol{\Phi}_{int}(t) + \boldsymbol{\Phi}_{frac}(t). \tag{2.5}$$

Assume two signals, *Ref* and *Div* whose cycles are defined by their respective rising edges. Moreover, we assume that the periods of *Ref* and *Div* are constant and variable with respect to time, respectively. This assumption is consistent with the operations of an ADPLL that we will analyse in the next sections.

By means of Eqs. (2.3) and (2.4), we can define the integer and fractional parts of the phase $\boldsymbol{\Phi}_{Ref}$ of *Ref* at the instants $t_{Div}[n]$ at which the edges of *Div* occur, as shown in Fig. 2.2. The integer part of $\boldsymbol{\Phi}_{Ref}$ at the instant $t_{Div}[n]$ is determined by the number of previous edges of *Ref*, where n is the index associated with the edges of *Div*. The fractional part of $\boldsymbol{\Phi}_{Ref}$ at $t_{Div}[n]$ is equal to $\Delta t[n]/T_{Ref}$, where $\Delta t[n]$ is the

Fig. 2.3 Phase of the signal *Div* expressed in cycles

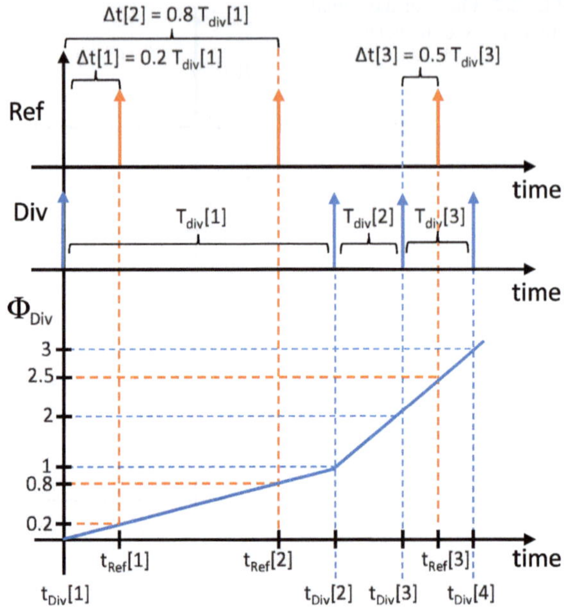

duration of the time interval between $t_{Div}[n]$ and the instant of the previous edge of *Ref*, as shown in Fig. 2.2.

It is important to notice from Eqs. (2.3) and (2.4) that we can also define the phase $\boldsymbol{\Phi}_{Div}$ of *Div* at the instants $t_{Ref}[n]$ at which the edges of *Ref* occur, as shown in Fig. 2.3.

The integer part of $\boldsymbol{\Phi}_{Div}$ at the instant $t_{Ref}[n]$ is equal to the number of the previous edges of *Ref*, where n is the index associated with the edges of *Ref*. The fractional part of $\boldsymbol{\Phi}_{Div}$ at $t_{Ref}[n]$ is equal to $\Delta t[n]/T_{Div}[k]$, where $\Delta t[n]$ is the duration of the time interval between $t_{Ref}[n]$ and the instant of the previous edge of *Div*, and $T_{Div}[k]$ is the duration of the k-th cycle of *Div* which includes $t_{Ref}[n]$, as shown in Fig. 2.3.

Finally, the phase difference between *Ref* and *Div* can be determined from Figs. 2.2 and 2.3, as shown in Fig. 2.4.

Considerations of the fractional and integer phases of *Ref* and *Div* and the definition of the phase difference between two signals are fundamental to understanding and comparing the operations of the ADPLL architectures that we will consider in the next sections.

2.3 Architectures of Phase-Difference Digitizing and Phase Digitizing ADPLLs

The operation of an ADPLL is based on the association of a digital word with the phase difference between two signals. In order to realize this association, it is possible to follow two approaches. The first approach is to produce an analog measurement

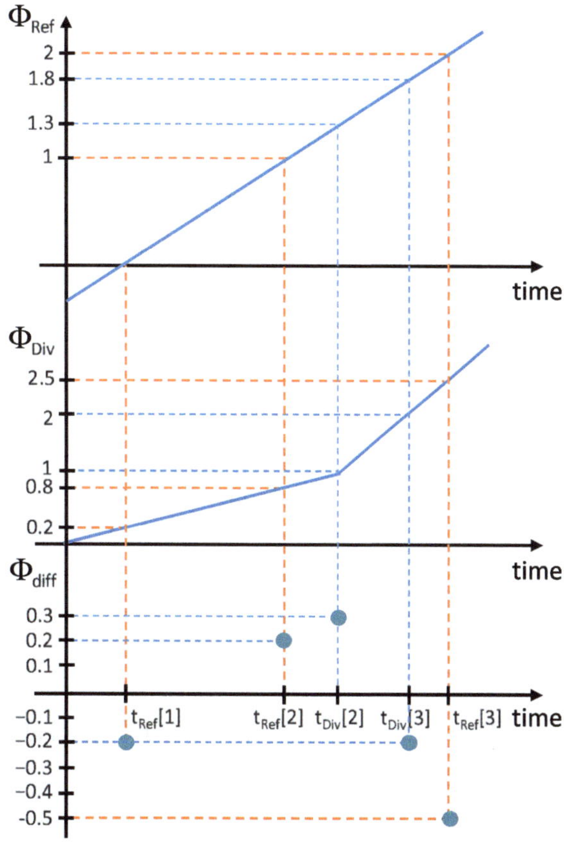

Fig. 2.4 Sampled phase difference between *Ref* and *Div* at time instants $t_{Ref}[i]$ and $t_{Div}[i]$

of the phase difference and then to convert the analog measurement to a digital word [1], as illustrated in Fig. 2.5a. ADPLLs that use this approach can be classified as "phase-difference digitizing" ADPLLs.

The second approach to realize an ADPLL is to produce digital words related to the phases of the signals and then to calculate the difference between the digitized phases [1], as shown in Fig. 2.5b. ADPLLs that use this approach can be classified as "phase digitizing" ADPLLs. The block diagram in Fig. 2.5a can be related to the operations of PFD-plus-TDC-based, TDC-based and flip-flop-based ADPLLs. The block diagram in Fig. 2.5b is associated with accumulator-based ADPLLs.

In the next sections, we will show how the integer and fractional parts of the phases of the signals in PFD-plus-TDC-based, TDC-based, flip-flop-based and accumulator-based ADPLL are converted to digital words.

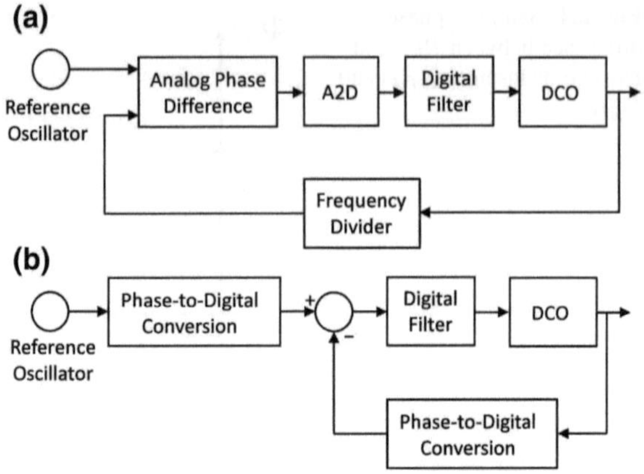

Fig. 2.5 ADPLL block diagrams in which **a** the phase difference is first associated with an analog signal and then converted to a digital word and **b** the phases of the reference oscillator and the DCO are converted to digital words and then subtracted

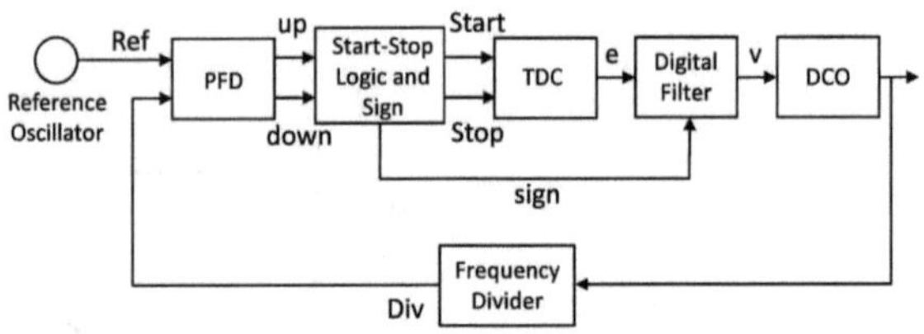

Fig. 2.6 Block diagram of a PFD-plus-TDC-based ADPLL

2.4 Phase Digitization in a PFD-Plus-TDC-Based ADPLL

The architecture of a PFD-plus-TDC-based ADPLL is shown in Fig. 2.6 [2–4]. Some works in the literature [2–4] do not show the block labelled "Start–Stop Logic and Sign" in Fig. 2.6. In these works, the operations of the block "Start–Stop Logic and Sign" are included in the TDC. We separate the block "Start–Stop Logic and Sign" from the TDC in order to simplify the description of the operation of a PFD-plus-TDC-based ADPLL.

In a PFD-plus-TDC-based ADPLL, the phase difference between the signals *Ref* and *Div* is measured by means of a Phase-Frequency-Detector (PFD).

Figure 2.7a, b shows the state diagram and the implementation with logic gates of a tristate PFD, respectively. The blocks labelled "DFF" in Fig. 2.7b represent D-flipflops.

In the PFD-plus-TDC-based ADPLL in Fig. 2.6, the block labelled "Start–Stop Logic" produces the *Start* and *Stop* signals for a TDC from the up and down pulses of a PDF.

Example signals *up* and *down* are shown in Fig. 2.8. We assume that the block "Start–Stop Logic" starts and stops the TDC when the PFD starts and stops producing a pulse, respectively. The signals "Start" and "Stop" are equal to "*up* OR *down*" and "(NOT *up*) AND (NOT *down*)", respectively, as shown in Fig. 2.8. The signal *sign*

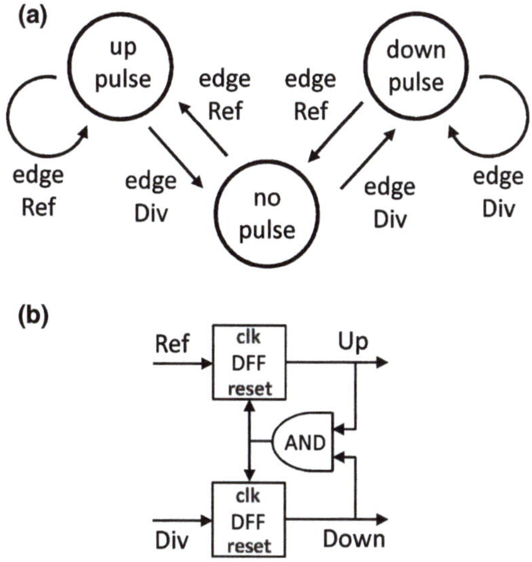

Fig. 2.7 **a** State diagram and **b** implementation with logic gates of a tristate PFD

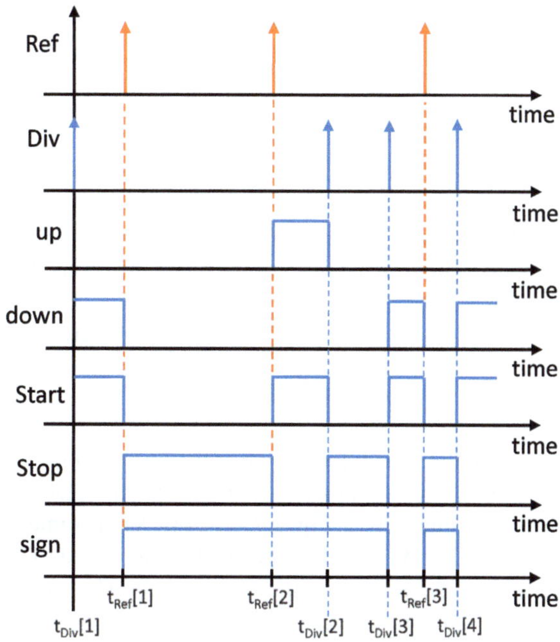

Fig. 2.8 Example timing diagrams showing the signals *up*, *down*, *Start*, *Stop* and *sign* in a PFD-plus-TDC-based ADPLL

is equal to "*up* OR ((NOT *up*) AND (NOT *down*))"; it is high during an up pulse and when the signals *up* and *down* are low at the same time. The signal *sign* is low during a down pulse.

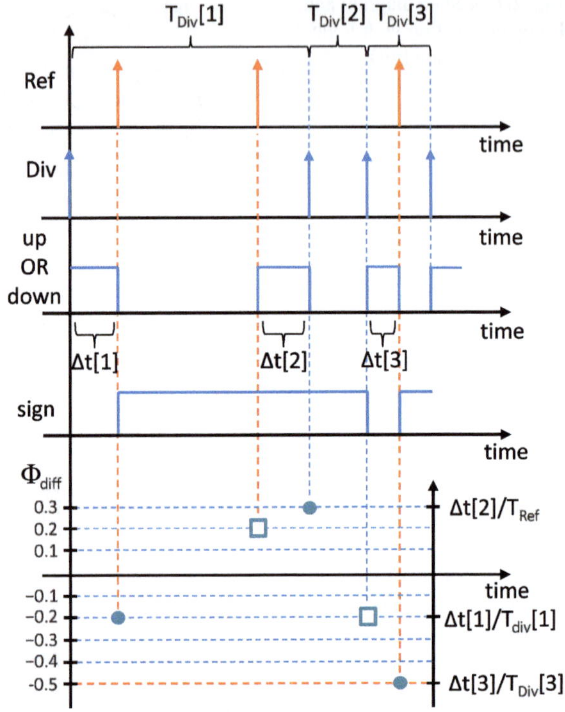

Fig. 2.9 Phase difference, state of the PFD and TDC output. The *circles* represents the measurement produced by the TDC combined with the signal *sign*

The durations of the up and down pulses that are produced by the PFD in Fig. 2.6 provide an *analog* measure of the phase difference between the signals *Ref* and *Div*, as shown in Fig. 2.9.

The signal *sign* is high when the phase difference is zero or positive, as shown in Fig. 2.9. The solid circles in Fig. 2.9 represent the measurements produced by the TDC combined with the signal *sign*.

The open squares in Fig. 2.9 are associated with values of the phase difference related to the time intervals when the *up* and *down* signals are both low. In order to keep our analysis simple, we assumed that the TDC does not measure the duration of these time intervals. In Fig. 2.9, we consider the same signals *Ref*, *Div* and $\boldsymbol{\Phi}_{diff}$ shown in Figs. 2.2, 2.3 and 2.4, respectively. The periods T_{Ref} and $T_{Div}[k]$ are constant and variable with respect to time, respectively. The index k in $T_{Div}[k]$ indicates the k-th cycle of *Div*. Moreover, we assume that the output of the TDC before normalization is a measure of $\Delta t[n]$ expressed in seconds, where n is the index associated with each pulse produced by the PFD. In order to obtain the phase difference between *Ref* and *Div*, the output of the TDC has to be normalized by T_{Ref} and $T_{Div}[k]$ when the *last edge* of the up or down pulse associated with $\Delta t[n]$ belongs to *Div* or *Ref*, respectively, according to the definitions given in Sect. 2.2 and as shown in Fig. 2.9. Hence, the output of the TDC in the architecture in Fig. 2.6 has to be normalized by a factor that alternates appropriately in time between T_{Ref} and $T_{Div}[k]$ in order to keep the phase measurement consistent. This important design consideration is *missing in the literature* on PFD-plus-TDC-based ADPLLs.

The sign of the phase difference is positive and negative, when the signal *sign* is high and low, respectively, as shown in Fig. 2.9. We assume that the signal *sign* is high when the signals *up* and *down* are low at the same time. However, the TDC produces measurements of the modulus of the phase difference only after a falling edge of an up or down pulse. Hence, the level of the signal *sign* is not related to the measurement of the phase difference when the signals *up* and *down* are low at the same time, and it can be set to low or high arbitrarily. Finally, notice that the durations of the up and down pulses in a PFD-plus-TDC-based ADPLL are related to the modulus of the phase difference including *both integer and fractional* parts.

In the example timing diagrams in Fig. 2.9, the phase difference between *Ref* and *Div* has an integer part equal to zero. The fact that the integer part of the phase difference is equal to or different from zero changes drastically how a PFD-plus-TDC based ADPLL operates. In the following, we will first explain the operations of an PFD-plus-TDC based ADPLL when the integer part of the phase difference is different from zero. Then, we will resume the analysis of a PFD-plus-TDC based ADPLL in which the integer part of the phase difference is *equal* to zero.

2.4.1 Phase Difference with Integer Part Different from Zero

The phase difference between *Ref* and *Div* has a non-zero integer part when one or more complete cycles of *Div* or *Ref* are included between the rising and falling edges of an up or down pulse, respectively. In this section, we explain the operations of a PFD-plus-TDC-based ADPLL that can measure the integer part of the phase difference. However, the ADPLLs that are described in [2, 4] assume that the phase difference always has an integer part that is equal to zero. In fact, the implementation of an ADPLL becomes simpler when the ADPLL can measure only small values of phase difference. Hence, we illustrate in the following the complications that arise in a PFD-plus-TDC-based ADPLL that measures the integer and fractional parts of the phase difference.

Assume that T_{sum} is the sum of the durations of all these cycles.[1] The fractional part of the phase difference $\boldsymbol{\Phi}_{diff}$ can be calculated by subtracting T_{sum} from the output of the TDC and then by normalizing the result, as explained in the initial part of this section. The integer part of the phase difference can be obtained by counting the number of complete cycles included in each up and down pulse. Let us illustrate this with an example.

Figure 2.10 shows example timing diagrams with the edges of the *Ref* and *Div*, and the respective phases $\boldsymbol{\Phi}_{Ref}$ and $\boldsymbol{\Phi}_{Div}$. The sampled phases of $\boldsymbol{\Phi}_{Ref}$ and $\boldsymbol{\Phi}_{Div}$ at the instants $t_{Ref}[i]$ and $t_{Div}[i]$ are associated with the dashed and solid curves in Fig. 2.10, respectively. Notice that $\boldsymbol{\Phi}_{Ref}$ and $\boldsymbol{\Phi}_{Div}$ are associated with two different scales. The

---

[1] The durations of the periods of T_{Div} can be estimated from the input of the DCO.

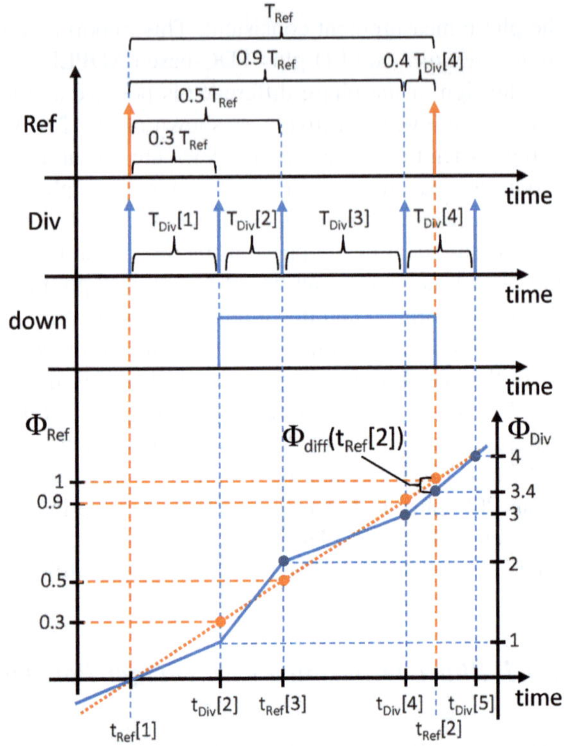

Fig. 2.10 Example timing diagram of the edges of *Ref* and *Div*

phase difference $\boldsymbol{\Phi}_{diff}$ between $\boldsymbol{\Phi}_{Ref}$ and $\boldsymbol{\Phi}_{Div}$ can be obtained by subtracting $\boldsymbol{\Phi}_{Div}$ from $\boldsymbol{\Phi}_{Ref}$. The phase difference $\boldsymbol{\Phi}_{diff}(t_{Ref}[2])$ is equal to -2.4 and, hence, it has a *non-zero* integer part. In Fig. 2.10, the time interval Δt that the TDC measures is equal to $T_{Div}[2] + T_{Div}[3] + 0.4T_{Div}[4]$. The fractional part of $\boldsymbol{\Phi}_{diff}(t_{Ref}[2])$ is given by $(\Delta t - T_{Div}[2] - T_{Div}[3])/T_{Div}[4]) = 0.4$. In a PFD-plus-TDC-based ADPLL the fractional part of $\boldsymbol{\Phi}_{diff}$ can be obtained by subtracting T_{sum} from the output of the TDC before normalizing. By contrast, the integer part of $\boldsymbol{\Phi}_{diff}(t_{Ref}[2])$ is given by the number of complete cycles of *Div* that are included between the rising and falling edges of the *down* pulse. In a PFD-plus-TDC based ADPLL, the integer part of $\boldsymbol{\Phi}_{diff}$ can be produced by a counter that counts the number of complete cycles included in the up and down pulses. The signal *e* in Fig. 2.6 can be obtained as the sum of the fractional and integer parts of the phase difference.

The digital filter in a PFD-plus-TDC-based ADPLL can be clocked on the falling edges of the up and down pulses. However, the duration of the time interval between consecutive edges of the down and up pulses is variable. Hence, the period of the clock of the digital filter is variable when the digital filter is clocked by the falling edges of the up and down pulses. This may compromise the operation of the digital filter, as we will discuss in Sect. 2.10.2. In order to keep the period of the clock of the digital filter constant over time, it is possible to clock the digital filter with

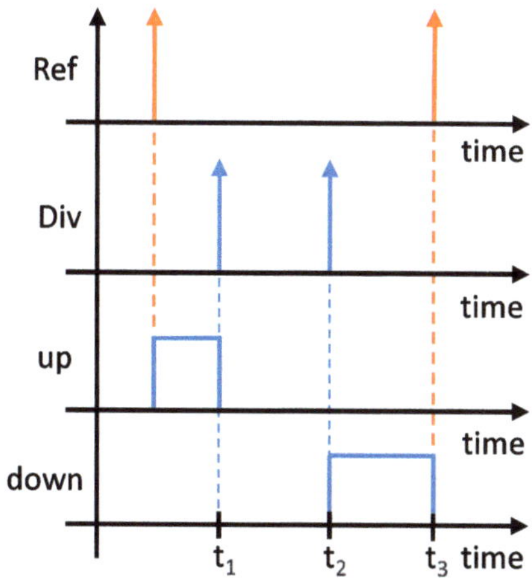

Fig. 2.11 Example pair of up and down pulses included between two consecutive edges of *Ref*

the output of the reference oscillator. However, when the digital filter is clocked by the reference oscillator, only the last measurement of the phase difference produced by the TDC during a cycle of the reference oscillator is processed by the digital filter. For example, Fig. 2.11 shows a pair up and down pulses included between two consecutive edges of *Ref*. When the configuration of pulses in Fig. 2.11 happens in a PFD-plus-TDC-based ADPLL, the TDC produces measurements of the phase difference at the instants t_1 and t_3 in Fig. 2.11.

When the digital filter is clocked by the falling edges of the up and down pulses, both the phase difference measurements at the instants t_1 and t_3 are sampled and processed by the digital filter. However, only the phase difference measurement at the instant t_3 is sampled and processed by the digital filter when the digital filter is clocked by the reference oscillator.

It is convenient to illustrate the phase-to-digital conversion in a PFD-plus-TDC-based ADPLL by means of a behavioral block diagram in order to be able to show the effects of the TDC quantization on the ADPLL's noise performance and to compare a PFD-plus-TDC-based ADPLL with the other ADPLL architectures. We will compare the main ADPLL architectures in terms of quantization error in Sect. 2.9.

In fact, the system that comprises the PFD, the block "Start–Stop Logic and Sign" and the TDC in Fig. 2.6 is a phase-to-digital converter that can be represented by the model in Fig. 2.12. The block labelled "Ideal Phase Difference Measurement" in Fig. 2.12 produces an ideal measurement of the phase difference between *Ref* and *Div* at the end of each up and down pulse. The block "Ideal Phase Difference Measurement" represents the operations of an ideal PFD. The phase difference is expressed in cycles of *Ref* or *Div* when the last edge of the up or down pulse belongs to *Div* or *Ref*, respectively. The block labelled "Modulus and Conversion to Seconds" in Fig. 2.12 multiplies the modulus of the output of the block "Ideal Phase Difference Measurement" by the period of *Ref* or *Div* when the last edge of the up or down pulse belongs to *Div* or *Ref*, respectively. Notice that two up or down pulses with

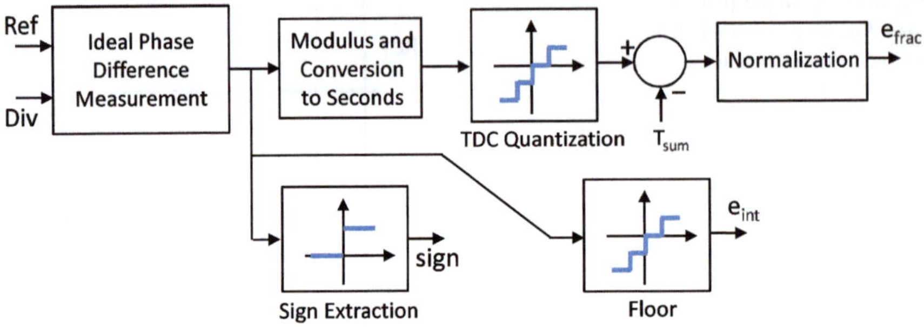

Fig. 2.12 Model of the system which comprises the PFD, the block "Start–Stop Logic and Sign" and the TDC in Fig. 2.6

different durations can produce the *same* phase difference at the output of the block "Ideal Phase Difference Measurement" when the last edges of the two pulses belong to *Ref* and *Div*, respectively. This is consistent with the fact that the same value of phase difference gives two different values at the output of the block "Conversion to Seconds" when the last edge of the up or down pulse belongs to *Div* or *Ref*.

The block labelled "TDC Quantization" in Fig. 2.12 represents the quantization effects of a TDC with finite time resolution *res*. The output *out* of the block "TDC Quantization" is given by:

$$out = res \cdot \text{floor}(in/res) \tag{2.6}$$

where floor$(\cdot)$ rounds to the largest integer smaller than the argument and *res* is the time resolution of the TDC. The block labelled "Normalization" in Fig. 2.12 divides its input by the period of *Ref* or *Div* when the last edge of the up or down pulse belongs to *Div* or *Ref*, respectively. Finally, the output of the block labelled "Sign Extraction" in Fig. 2.12 is equal to one when its input is positive or zero; otherwise it is zero.

The model in Fig. 2.12 can be modified to show the effect of a wrong normalization of the TDC output on the phase measurement in a PFD-plus-TDC-based ADPLL. We illustrate this effect by means of an example. Assume an example PFD-plus-TDC-based ADPLL in which the output of the TDC is normalized by T_{Ref} independently of the last edge of the up and down pulses. Moreover, we assume that the quantization in the TDC is negligible in order to keep our analysis simple. The block diagram in Fig. 2.12 simplifies to that in Fig. 2.13 when the output of the TDC is always normalized to the period T_{Ref} of *Ref* and the quantization noise in the TDC is negligible.

In the block diagram in Fig. 2.13, the signal *e*, which is the TDC output after normalization, is equal to the phase difference when the phase difference is positive. By contrast, the signal *e* in Fig. 2.13 is equal to $T_{Ref}/T_{Div}[k]$ times the phase difference when the phase difference is negative. The characteristic of the phase-to-digital conversion can be defined by the normalized TDC output as a function of the input

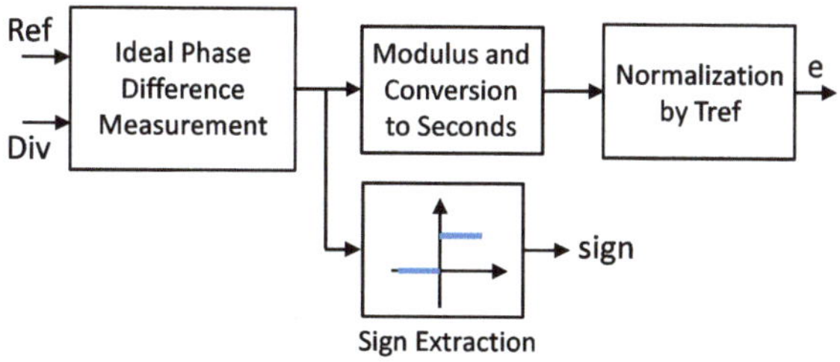

Fig. 2.13 Block diagram of the phase-to-digital conversion in a PFD-plus-TDC-based ADPLL with the output of the TDC normalized always by T_{Ref}

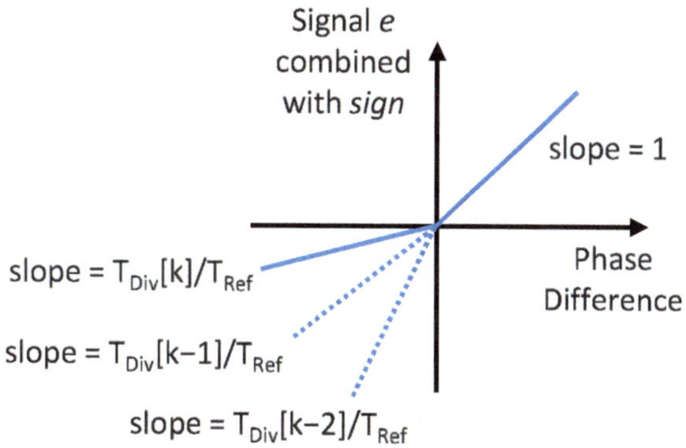

Fig. 2.14 Example characteristic of the phase-to-digital conversion associated with the block diagram in Fig. 2.13

phase difference. The TDC characteristic of the phase-to-digital conversion of the block diagram in Fig. 2.13 has a time-varying slope for negative values of the phase difference, as shown in Fig. 2.14.

The characteristic in Fig. 2.14 is nonlinear and time-varying. When the slope of the phase-to-digital characteristic is time-varying, the gain of the phase-to-digital characteristic changes over time. A time-varying gain in an ADPLL has to be taken into account in the design of the ADPLL in order to guarantee that the stability requirements are met for all possible values of the gain. Moreover, an ADPLL with a characteristic of the type in Fig. 2.14 shows different transient responses to positive and negative steps in the phase difference.

In order to avoid the nonlinear time-varying phase-to-digital characteristic in Fig. 2.14, it is necessary to normalize the output of the TDC by the period of *Ref* or *Div* when the the falling edge of an up or down pulse is associated with an edge of *Div* or *Ref*, respectively. Hence, the periods of *Ref* and *Div* have to be known in order to avoid this nonlinear time-varying characteristic. Moreover, the calculation of T_{sum} in Fig. 2.12 requires that all complete periods of *Div* included in an up or down pulse are known. Unfortunately, the period of *Div* is not always known, so this nonlinear time-varying effect cannot be eliminated by normalizing the TDC output.

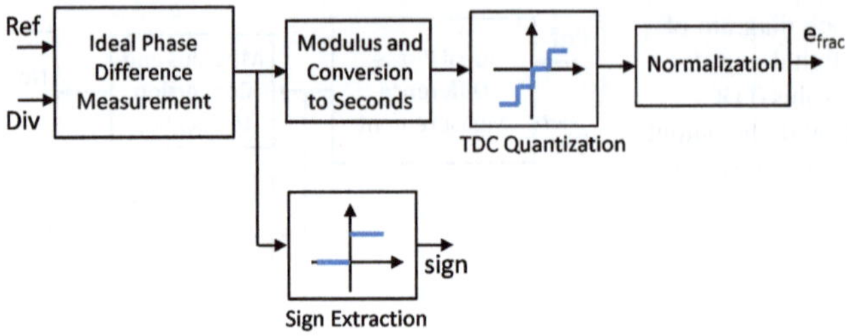

Fig. 2.15 Model of the system which comprises the PFD, the block "Start–Stop Logic and Sign" and the TDC in Fig. 2.6 when the phase difference has integer part equal to zero

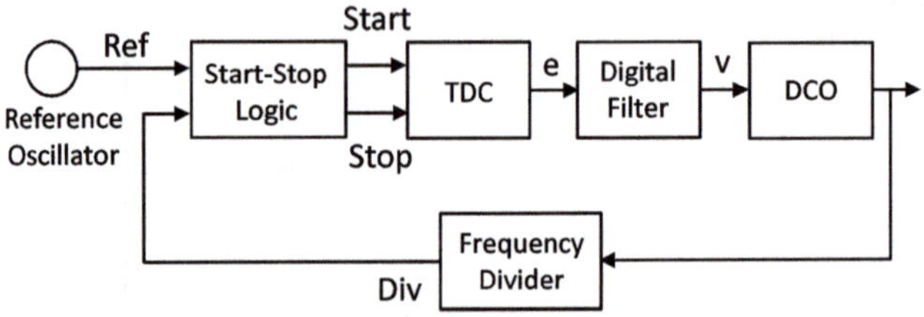

Fig. 2.16 Block diagram of a TDC-based ADPLL

2.4.2 Phase Difference with Integer Part Equal to Zero

The integer part of the phase difference is zero when there are one or two edges of *Div* between two consecutive edges of *Ref*. When there are one or two edges of *Div* between two consecutive edges of *Ref*, the up and down pulses produced by the PFD never include a complete period of *Ref* or *Div*. Hence, T_{sum} in the model in Fig. 2.12 is always equal to zero. The normalized output of the TDC in a PFD-plus-TDC-based ADPLL is equal to the phase difference when the integer part of the phase difference is zero. The model in Fig. 2.12 simplifies to that in Fig. 2.15. Hence, it is not necessary to subtract T_{sum} from the output of the TDC in a PFD-plus-TDC-based ADPLL in which the integer part of the phase difference is zero.

2.5 Phase Digitization in TDC-Based ADPLL

Figure 2.16 shows the block diagram of a TDC-based ADPLL [5]. The TDC in Fig. 2.16 starts and stops a measurement on a rising edge of *Ref* and on the next edge of *Div* [5], respectively. Hence, the output of the TDC normalized by T_{Ref} is equal to the fractional part of the phase of the reference oscillator on the edges of *Div*. The fractional part of the phase of the reference oscillator is equal to the fractional part of the phase difference Φ_{Ref} measured on the edges of *Div* when the integer parts

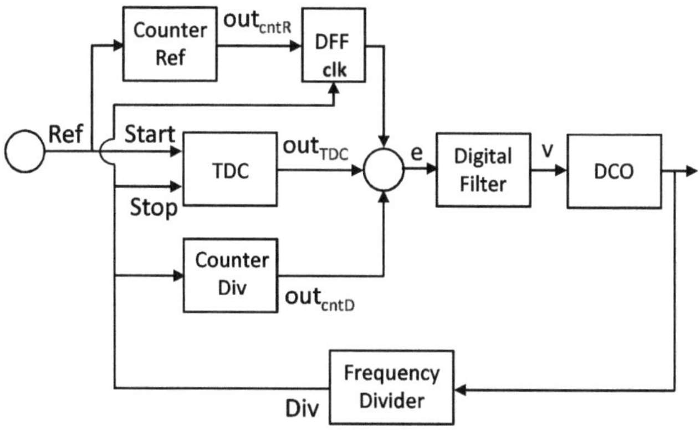

Fig. 2.17 Block diagram of a TDC-based ADPLL with counters to measure integer and fractional parts of the phase difference

of the phases of *Ref* and *Div* are equal on the edges of *Div*. Hence, the TDC-based ADPLL in Fig. 2.16 does not measure the integer part of the phase difference.

The digital filter in a TDC-based ADPLL can be clocked by the *Stop* signal in order to let the digital filter process the phase difference measurements as soon as the TDC produces them. The duration of the time interval between two consecutive edges of the *Stop* signal is varying over time. This may also lead to problems in the operation of the digital filter, as we will discuss in Sect. 2.10.2. By contrast, the period of the clock of the digital filter is constant when the digital filter is clocked by the reference oscillator.

Before analysing the architecture in Fig. 2.16, it is important to explain the operations of a TDC-based ADPLL that can measure the *integer and fractional* parts of the phase difference. The TDC-based ADPLL that we will first describe can measure the phase difference when the integer part of the phase difference is different from zero. Then, we will resume our analysis of the TDC-based ADPLL in Fig. 2.16 that can measure the phase difference only when the integer part of the phase difference is equal to zero.

2.5.1 Phase Difference with Integer Part Different from Zero

Two counters can be introduced in the ADPLL in Fig. 2.16 in order to measure the integer part of the phase difference, as shown in Fig. 2.17. The ADPLL in Fig. 2.17 measures the phase difference between *Ref* and *Div* at the instant associated with the first edge of *Div* that occurs after an edge of *Ref*. In order to explain how the phase difference is digitized in the ADPLL in Fig. 2.17, we consider the example timing diagrams of the edges of *Ref* and *Div* and their phases, as shown in Fig. 2.18.

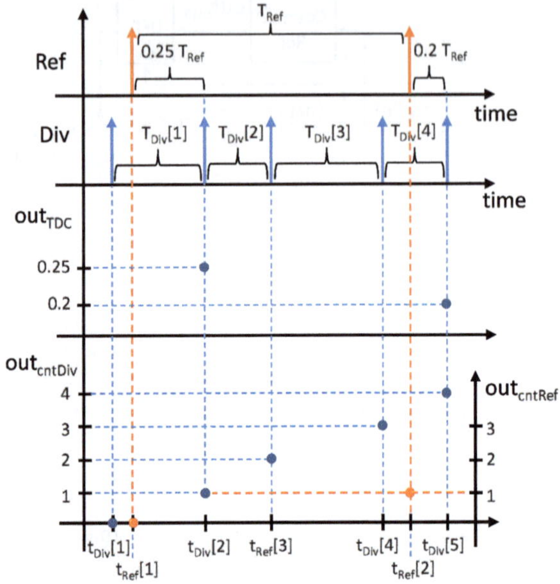

Fig. 2.18 Example timing diagrams of the edges of *Ref* and *Div* and the outputs of the TDC and the counters in the ADPLL in Fig. 2.17

Notice that the integer part of the difference between the example phases of *Ref* and *Div* in Fig. 2.18 is different from zero. Moreover, the instants associated with the first edges of *Div* that occur after the edges of *Ref* are $t_{Div}[2]$ and $t_{Div}[5]$, respectively. The phase differences $\boldsymbol{\Phi}_{diff}(t_{Div}[2])$ and $\boldsymbol{\Phi}_{diff}(t_{Div}[5])$ at the instants $t_{Div}[2]$ and $t_{Div}[5]$ are equal to -0.75 and to -2.8, respectively. The outputs out_{cntRef} and out_{cntDiv} of the counters in Fig. 2.17 are equal to the integer parts of the phases of *Ref* and *Div*, respectively, as shown in Fig. 2.18. The phase of *Ref* with integer and fractional parts is obtained by adding out_{cntRef} and out_{TDC}. The phase of *Div* is given by out_{cntDiv}. Hence, a digital measurement e of the phase difference is given by:

$$e = out_{cntRef} + out_{TDC} - out_{cntDiv}. \tag{2.7}$$

By applying Eq. (2.7) to the example in Fig. 2.18, the digitized phase differences $e(t_{Div}[2])$ and $e(t_{Div}[5])$ are equal to -0.75 and to -2.8, respectively, as expected.

The phase-to-digital conversion realized by the TDC in the ADPLL in Fig. 2.17 is represented by the block diagram in Fig. 2.19. The output of the block labelled "Ideal Phase Measurement on the Edges of Div" is equal to the phase of its input on the edges of *Div*. The output of the block labelled "Modulus and Conversion to Seconds" is equal to the modulus of the input multiplied by T_{Ref}. The blocks labelled "TDC Quantization" and "Normalization by T_{Ref}" represent the quantization in the TDC and a division by T_{Ref}, respectively.

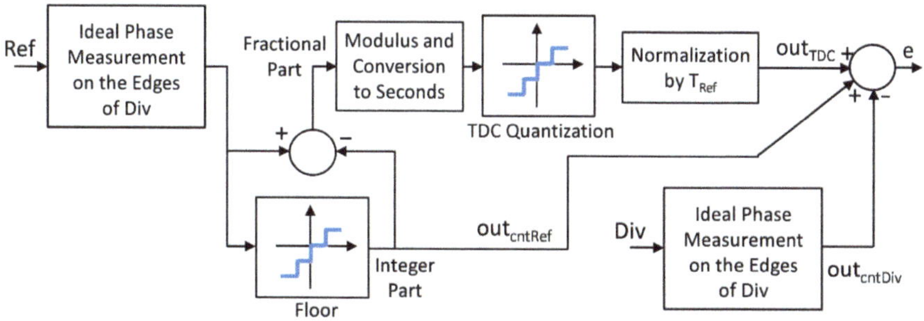

Fig. 2.19 Model of the phase digitization realized by system that comprises the block TDC and the counters in Fig. 2.17

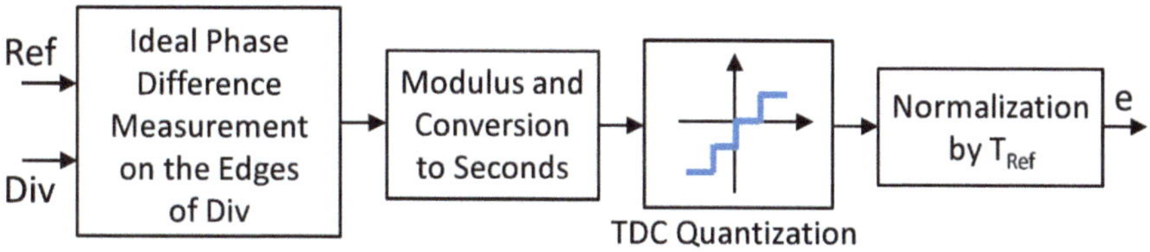

Fig. 2.20 Model of the phase digitization realized by the TDC in Fig. 2.16

2.5.2 Phase Difference with Integer Part Equal to Zero

The output of the TDC in the TDC-based ADPLL in Fig. 2.16 is a digitized version of the phase difference between *Ref* and *Div* when the integer part of the phase difference is equal to zero. Moreover, the TDC can only measure the *positive* values of the fractional part of the phase difference. Hence, it is important to introduce a fixed delay between *Ref* and *Div* in order to guarantee that the fractional part of the phase error is always positive. For example, a fixed delay between *Ref* and *Div* can be set by introducing a delay element (such as a non-inverting buffer) with delay t_{offset} at one of the TDC inputs and subtracting from the TDC output a constant which is equal to the output produced by the TDC when the time interval that has to be measured is equal to t_{offset}. In fact, an ideal TDC in a noiseless TDC-based ADPLL in this configuration produces an output equal to t_{offset} when the ADPLL has reached steady state. The input of the digital filter in this ADPLL is given by the difference between the TDC output and t_{offset}. Hence, the input of the digital filter is equal to zero and the DCO frequency remains constant. As a consequence, the phase difference between *Ref* and *Div* is constant and equal to t_{offset}.

The phase-to-digital conversion realized by the TDC in a TDC-based ADPLL is represented by the block diagram in Fig. 2.20 when the integer part of the phase difference is equal to zero. The output of the block labelled "Ideal Phase Difference Measurement on the edges of Div" is equal to the phase difference between *Ref* and *Div*. The output of the block "Ideal Phase Difference Measurement on the edges

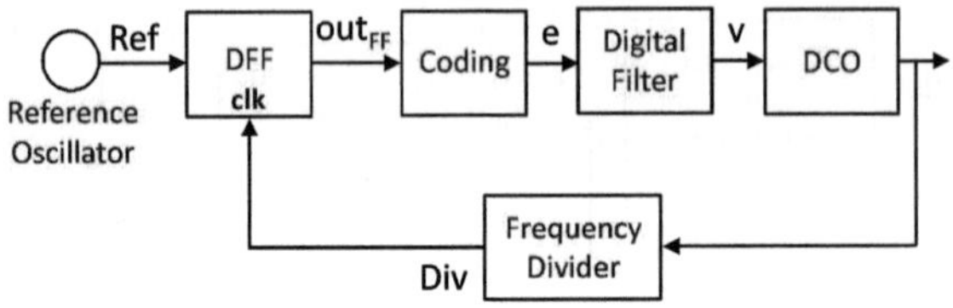

Fig. 2.21 Block diagram of a flip-flop-based ADPLL

of Div" is equal to the block "Ideal Phase Measurement on the edges of Div" with input *Ref* in Fig. 2.19 when the integer part of the phase difference is equal to zero. The blocks "Modulus and Conversion to Seconds", "TDC Quantization" and "Normalization by T_{Ref}" in Fig. 2.20 realize the same operations as the blocks with the same names in Fig. 2.19.

2.6 Phase Digitization in a Flip-Flop-Based ADPLL

A particular case of a TDC-based ADPLL is the flip-flop based ADPLL [6, 7] shown in Fig. 2.21.

When the integer part of the phase difference is zero, the output of the flip-flop in Fig. 2.21 is equal to 1 if a rising edge of *Ref* occurred before that of *Div*; otherwise it is zero. Hence, the output of the TDC is equal to zero and one when the phase difference on the edges of *Div* is negative and positive, respectively, assuming that the integer part of the phase difference in zero. When the integer part of the phase difference is different from zero, the level of the output of the flip-flop is not determined by the sign of the phase difference. For example, assume the example timing diagram in Fig. 2.22.

In Fig. 2.22, the period of *Div* is much smaller than that of *Ref* and the integer part of the phase difference is zero or negative. The output of the flip-flop is equal to one and zero when the edges of *Div* are included in the first and second halves of the period of *Ref*, respectively. Hence, the output of the flip-flop is not unique when the sign and the integer part of the phase difference are fixed and different from zero, respectively. We conclude that it is necessary that the integer part of the phase difference is always equal to zero in the flip-flop-based ADPLL shown in Fig. 2.21 in order to have a bilinear relationship between the output of the flip-flop and the sign of the phase difference.

The output of the block labelled "Coding" in Fig. 2.21 is the digital word $-wrd$ or $+wrd$ when the input is zero or one, respectively. The "Coding" block is necessary in order to increase or decrease the DCO frequency when the phase difference is positive or negative, respectively. In fact, if the output of the flip-flop in Fig. 2.22 is fed to the input of the digital filter directly, the input of the digital filter is zero when the phase difference between *Ref* and *Div* is negative. The output of a generic digital

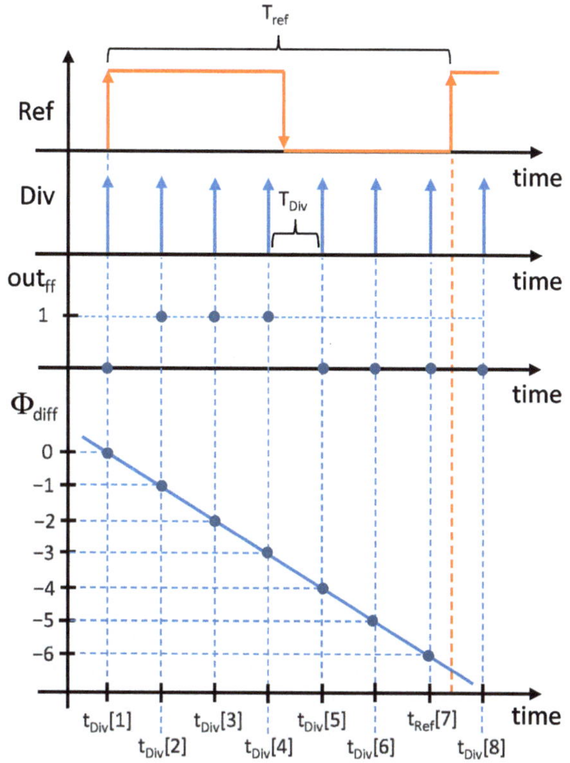

Fig. 2.22 Example timing diagram of the edges of *Ref* and *Div*, the phase difference between *Ref* and *Div* and the output out_{FF} of the flip-flop in Fig. 2.21, when the integer part of the phase difference is zero or negative

filter with one pole at the origin and several zeros and poles in the left half-plane[2] does not change when the input is zero. Hence, the ADPLL in Fig. 2.21 does not change the frequency or phase of the DCO when the phase difference between *Ref* and *Div* is zero and the D-flip-flop is connected to the digital filter directly. By contrast, the output of the "Coding" block in Fig. 2.21 is negative when the phase difference is negative. Thus, the DCO frequency decreases when the phase difference between *Ref* and *Div* is negative. The modulus of the digital words produced by the "Coding" block is equivalent to the gain of the TDC in the TDC-based ADPLL in Fig. 2.16.

Notice that it is important to determine when the phase measurements are produced in a flip-flop-based ADPLL in order to determine how the digital filter is clocked. In the ADPLL in Fig. 2.21, the measurement of the phase difference is produced on the edges of *Div*. Hence, we assume that the digital filter in a flip-flop-based ADPLL is clocked by *Div*.

Figure 2.23 shows a behavioral model associated with the phase-to-digital conversion in a flip-flop-based ADPLL. The block labelled "Ideal Phase Difference Measurement on the Edges of Div" in Fig. 2.23 produces an ideal measurement of

[2] We assume a generic digital filter with one pole in the origin in order to have a type-II PLL. Moreover, additional zeros and poles have to be in the left hand plane to guarantee a stable loop.

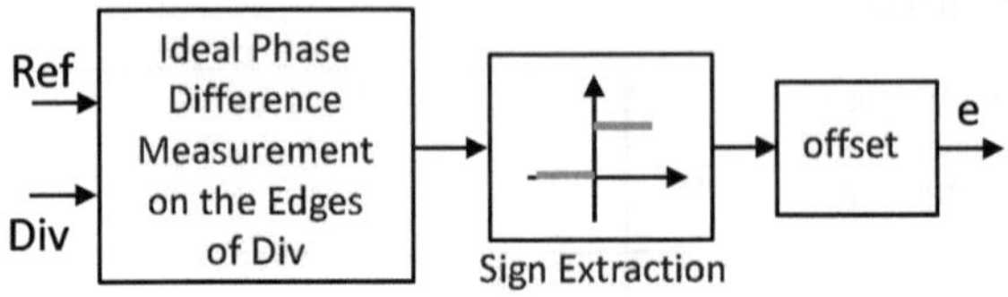

Fig. 2.23 Model of the phase-to-digital conversion associated with the flip-flop in the flip-flop-based ADPLL in Fig. 2.21 when the integer part of the phase difference is equal to zero

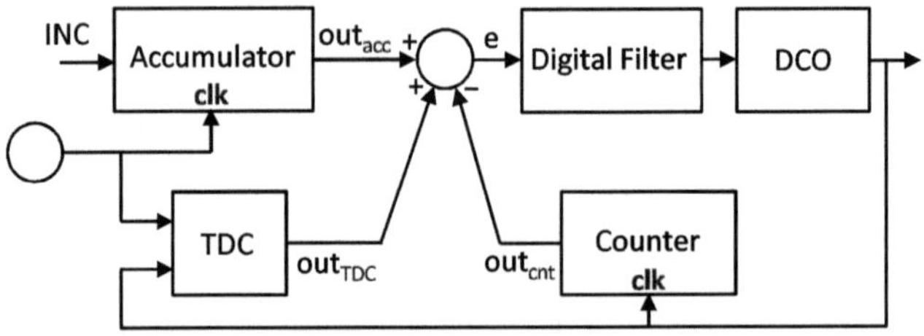

Fig. 2.24 Block diagram of an accumulator-based ADPLL

the phase difference between *Ref* and *Div* on the rising edges of *Div*. The block "Ideal Phase Difference Measurement" represents the operations of an ideal flip-flop. The output of the block labelled "Sign Extraction" in Fig. 2.23 is equal to 1 when the input is positive or zero. Otherwise, the output of the block "Sign Extraction" is zero. The main difference between a flip-flop-based and a TDC-plus-PFD based ADPLL is that a flip-flop-based ADPLL does not produce a measurement of the amplitude of the modulus of the phase difference. In fact, a flip-flop-based ADPLL discriminates only between the cases when the phase difference is positive or negative. Hence, the phase difference between *Ref* and *Div* cannot be set to a constant in an ideal noiseless flip-flop-based ADPLL at steady state. By contrast, in ideal noiseless PFD-plus-TDC-based and TDC-based ADPLLs the phase difference between *Ref* and *Div* can be a constant when the ADPLL has reached steady state. However, a flip-flop-based ADPLL requires only one flip-flop to produce the phase-to-digital conversion.

2.7 Phase Digitalization in an Accumulator-Based ADPLL

Figure 2.24 shows the block diagram of an accumulator-based ADPLL [8, 9].

The phase-to-digital conversion in the accumulator-based ADPLL in Fig. 2.24 can be studied in two cases: when the integer part of the difference between INC times the phase of the reference oscillator and that of the DCO is (i) different from zero and (ii) equal to zero. Notice that we will use the term "phase difference" in association with accumulator-based ADPLLs to refer to the difference between *INC*

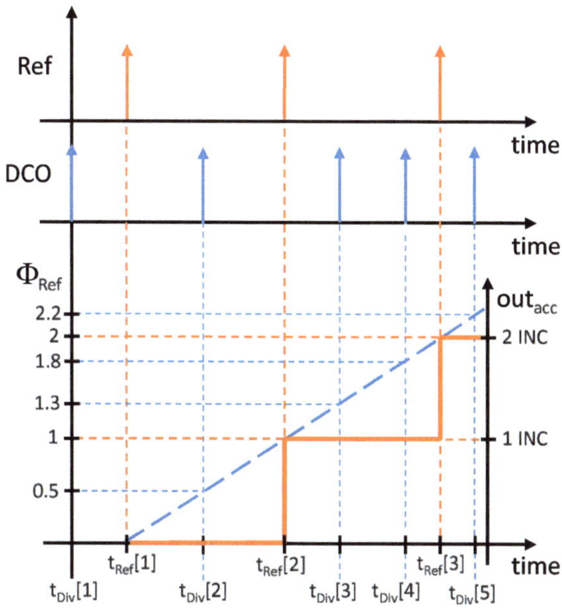

Fig. 2.25 Phase of the reference oscillator and output of the accumulator in an accumulator-based ADPLL

times the phase of the reference oscillator and that of the DCO. In the following, we will focus first on the operation of an accumulator-based ADPLL when the integer part of the phase difference is different from zero. Then, we will consider the case in which the integer part of the phase difference is *equal* to zero.

2.7.1 Phase Difference with Integer Part Different from Zero

Before discussing the operation of an accumulator-based ADPLL, notice that the accumulator and the counter in Fig. 2.24 operate in the clock domains of the reference oscillator and the DCO, respectively. Thus, in order to realize subtraction between the outputs of the accumulator and the counter, it is necessary to sample the output of the accumulator on the edges of the DCO or to sample the output of the counter on those of *Ref*. This sampling operation is referred to as "retiming" in most papers [8, 9]. In order to keep our analysis simple, we first discuss the basic operations of an ADPLL without retiming. Then, we will analyse the impact of retiming on the ADPLL.

First, let us assume that there is at least one edge of the DCO between two consecutive edges of the reference oscillator. This assumption is consistent with the fact that the frequency of the DCO is usually much higher than that of the reference oscillator.

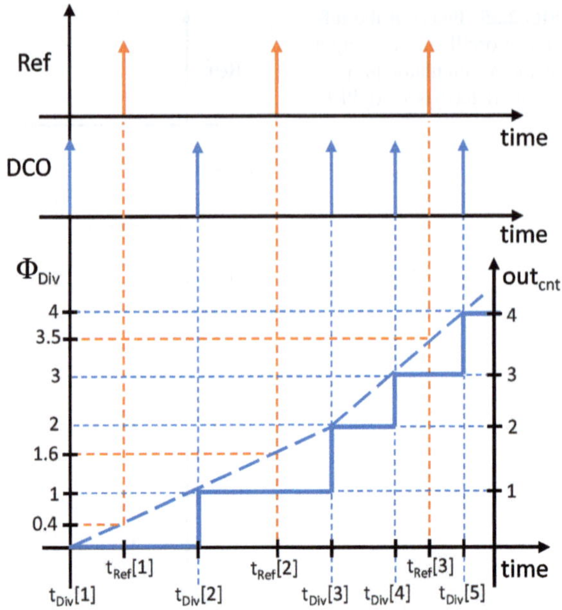

Fig. 2.26 Phase of the DCO and output of the counter in an accumulator-based ADPLL

The output of the accumulator in Fig. 2.24 is equal to the integer part of the phase of *Ref* multiplied by *INC*, according to the definition of phase that we used in Sect. 2.2. Figure 2.25 shows the relationship between the output of the accumulator in Fig. 2.24 and the phase of *Ref*. Notice that there is no frequency divider in the accumulator-based ADPLL in Fig. 2.24. The frequency divider is not needed because the ratio between the frequencies of the reference oscillator and the DCO is set by *INC*. In fact, if the phase difference at steady state is zero the phase of the DCO is equal to *INC* times that of the reference oscillator. Phase is the integral of frequency. Hence, the frequency of the DCO at steady state is equal to *INC* times that of the reference oscillator.

Similar to the accumulator, the output of the counter is equal to the integer part of the phase of the DCO, as shown in Fig. 2.26.

Let us assume that the time interval measured by the TDC in Fig. 2.24 is determined by an edge of *Ref* and the successive *DCO* edge. From the definitions of phase in Sect. 2.2, we deduce that the time interval measured by the TDC in Fig. 2.24 divided by T_{Ref} is a measure of the *fractional part* of the phase of the *reference oscillator* at the instants of the DCO edges, as illustrated in Fig. 2.27. Hence, the output of the TDC has to be *added* to the output of the accumulator. Notice that if the TDC is reset and started on the edges of the DCO and stopped on those of *Ref*, out_{TDC} is related to the fractional part of *the phase of the DCO* and the normalization factor of the TDC has to be the period of the DCO.

The phase difference in Fig. 2.24 is produced by a digital adder with three inputs, out_{acc}, out_{TDC} and out_{cnt}. The inputs to the digital adder are associated with *two*

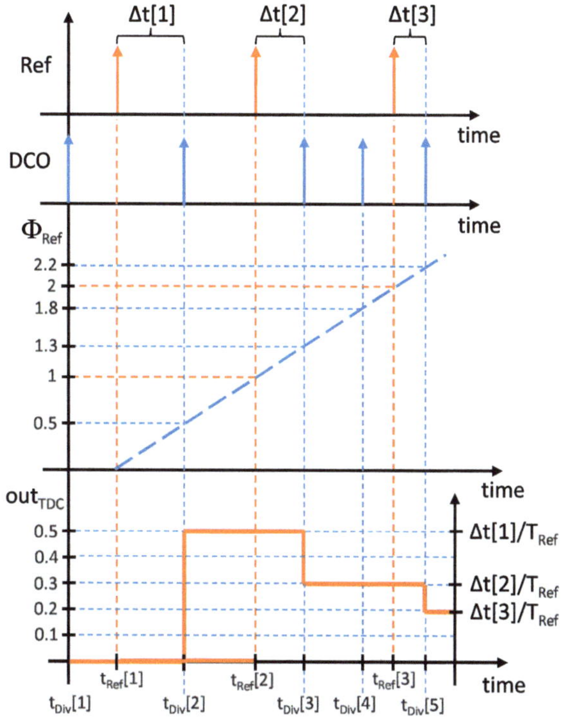

Fig. 2.27 Phase of the reference oscillator and output of the TDC in an accumulator-based ADPLL

different time domains. In fact, out_{acc} is associated with the time domain of the edges of *Ref*. By contrast, the signals out_{TDC} and out_{cnt} are related to the time domain of the edges of the DCO. In order to produce the addition in a single time domain, the inputs of the digital adder have to be sampled. By retiming the reference oscillator in an accumulator-based ADPLL, it is possible to have the signals out_{acc}, out_{TDC} and out_{cnt} in the same time domain.

Figure 2.28, shows an accumulator-based ADPLL with retiming [9]. The blocks labelled "DFF" in Fig. 2.28 which are connected to the outputs of the counter and the TDC represent arrays of flip-flops that sample the bits of the counter and TDC outputs, respectively. Retiming does not change the fact that the outputs of the accumulator and the counter in Fig. 2.28 are equal to the integer parts of the phases of the reference oscillator and the DCO on the edges of the output Ref_R of the block "Retiming". In fact, the output of the accumulator in Fig. 2.28 on the edges of Ref_R is equal to the output of an accumulator clocked by *Ref* and sampled on the edges of Ref_R. The TDC in Fig. 2.28 measures the fractional part of the phase of the reference oscillator on the edges of Ref_R. The TDC in Fig. 2.28 is started and stopped on the edges of *Ref* and the next edges of the DCO, respectively. Hence, the normalization factor of the TDC output is the period of the reference oscillator. Different from a PFD-plus-TDC based ADPLL, the normalization factor of the TDC in an accumulator-based ADPLL with retiming does not have to be changed over time because the TDC is

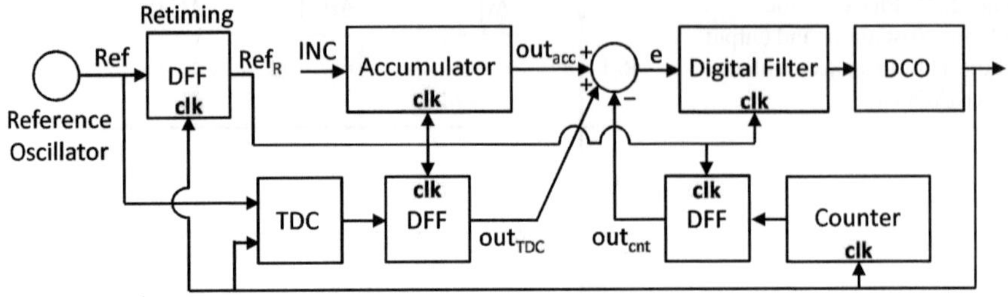

Fig. 2.28 Block diagram of an accumulator-based ADPLL with retiming based on sampling the output of the reference oscillator on the edges of the DCO

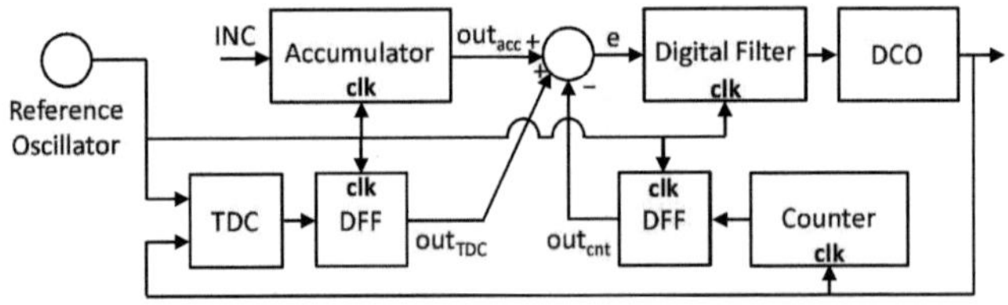

Fig. 2.29 Block diagram of an accumulator-based ADPLL with retiming based on the edges of the reference oscillator

always stopped on the edges of the same signal. Notice that the output of the TDC has to be multiplied by *INC* in order to have the measurement of the fractional part of the phase of *Ref* consistent with the measurement of the integer part produced by the accumulator.

The loop filter in Fig. 2.28 is clocked by the output Ref_R of the block "Retiming". The signal Ref_R has a variable period because it is obtained by sampling the output reference oscillator with the DCO output. In Sect. 2.10.2, we will discuss the effect of a variable period of the clock of the digital filter.

An alternative approach to retiming in Fig. 2.28 is shown in Fig. 2.29 [10]. In the approach in Fig. 2.29, the clock of the digital filter has a constant period and issues related to a noisy clock are thereby avoided in the digital filter [10]. The TDC in Fig. 2.29 is reset and started on the edges of the DCO and stopped on those of the reference oscillator [10].

The phase-to-digital conversion in an accumulator-based ADPLL with retiming in Fig. 2.28 is represented by the block diagram in Fig. 2.30.

The outputs of the blocks labelled "Ideal Phase Measurement on the Edges of Ref_R" in Fig. 2.30 are equal to the phases of their respective input signals measured on the edges of Ref_R. The output of the blocks labelled "Scaling by INC" in Fig. 2.30 are equal to its input times *INC*. The output of the block "Modulus and Conversion to Seconds" in Fig. 2.30 is equal to the modulus of its input times T_{Ref}. The block "Normalization by T_{Ref}" in Fig. 2.30 divides its input by T_{Ref}. The operations of the blocks labelled "Floor" and "TDC Quantization" in Fig. 2.30 are equal to those of the blocks with the same names in Fig. 2.17.

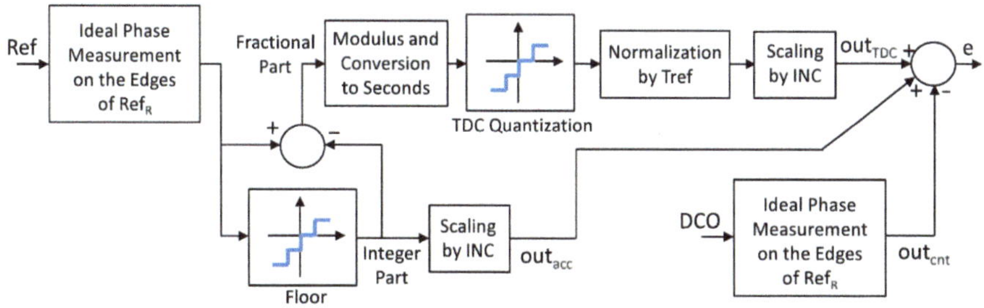

Fig. 2.30 Block diagram of an the phase-to-digital conversion in an accumulator-based ADPLL in Fig. 2.28

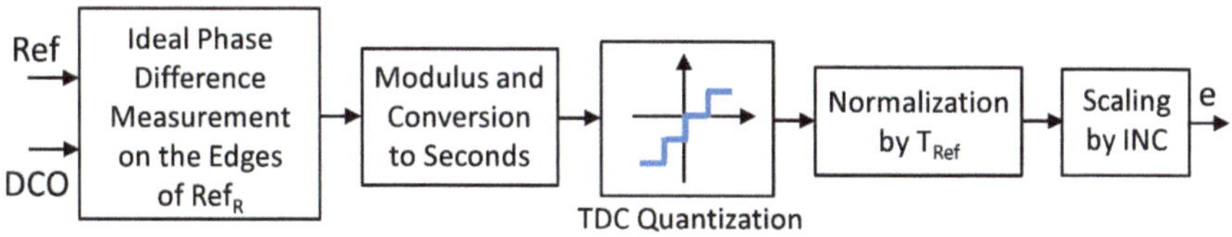

Fig. 2.31 Block diagram of the phase-to-digital conversion in an accumulator-based ADPLL in Fig. 2.28 when the integer part of the phase difference is equal to zero

The phase of the DCO in Fig. 2.28 on each edge of Ref_R is an integer. Hence, the output out_{cnt} of the counter in Fig. 2.28 on an edge of Ref_R is equal to the corresponding phase of the DCO. Thus, the output of the block "Ideal Phase Measurement on the Edges of Ref$_R$" associated with the DCO in Fig. 2.30 is equal to out_{cnt}.

In an accumulator-based ADPLL, the fractional part of the phase of *Ref* is quantized by the TDC. By contrast, in a PFD-plus-TDC-based ADPLL, the phase difference between *Ref* and *Div* with *both integer and fractional* parts is quantized by the TDC when the integer part of the phase difference is different from zero. This is the main difference between the phase-to-difference conversion in a PFD-plus-TDC-based and an accumulator-based ADPLL.

2.7.2 Phase Difference with Integer Part Equal to Zero

The integer part of the phase difference in the accumulator based ADPLL in Fig. 2.29 is equal to zero when the number of the DCO's edges included between two consecutive edges of the reference oscillator is equal to *INC*.

Assume that the time interval between the edges of *Ref* and the closest DCO edges is very small. Hence, the fractional part of the phase difference is not zero. Moreover, the number of complete DCO periods included between two consecutive edges of *Ref* is equal to ($INC - 1$). Hence, on every edge of Ref_R in Fig. 2.29, the outputs of the

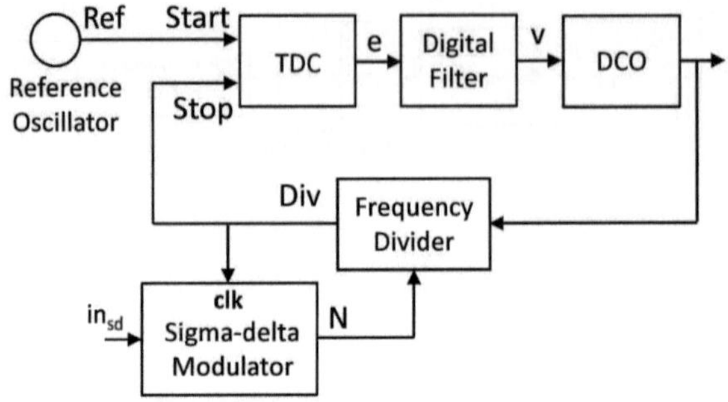

Fig. 2.32 Fractional-N TDC-based ADPLL with a sigma-delta modulator that sets the division ratio N

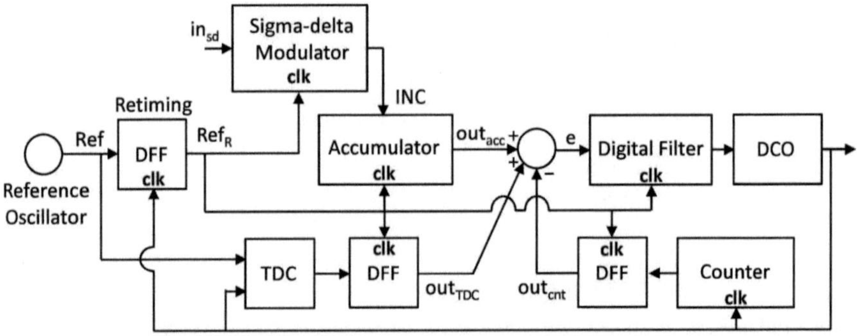

Fig. 2.33 Accumulator-based ADPLL with a sigma-delta modulator that sets the division ratio INC

counter and the accumulator are equal when the integer part of the phase difference is zero. By contrast, the fractional part of the phase difference is equal to the fractional part of the phase of *Ref*, i.e. the output of the TDC in Fig. 2.29. We conclude that the block diagram in Fig. 2.30 can be simplified to that in Fig. 2.31 when the integer part of the phase difference is zero in an accumulator-based ADPLL.

2.8 Fractional-N Operations

In traditional fractional-N PLLs, a sigma-delta modulator is used to set the division ratio of the frequency divider. Similar to a traditional PLL, a TDC-based ADPLL can include a sigma-delta modulator whose output determines the division ratio N, as shown in Fig. 2.32.

A similar approach can be used in accumulator-based ADPLLs.

Figure 2.33 shows an accumulator-based ADPLL with a sigma-delta modulator that sets the value of *INC*. To our knowledge, the architecture in Fig. 2.33 is miss-

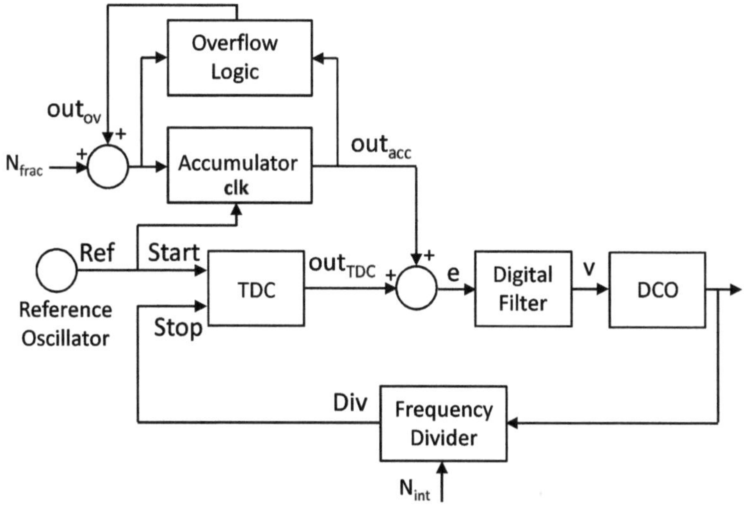

Fig. 2.34 Fractional-N TDC-based ADPLL with an accumulator that sets the fractional part N_{frac} of the division ratio

ing in the literature. By contrast, an accumulator-based ADPLL which synthesizes fractional division ratios is obtained by introducing a fractional part in the digital word *INC* such that the accumulator in the accumulator-based ADPLL accumulates $INC = INC_{int} + INC_{frac}$.

A similar approach can be applied to a TDC-based ADPLL by introducing an accumulator that increases by INC_{int} every time an edge of *Ref* occurs, as shown in Fig. 2.34.

The accumulator in Fig. 2.34 determines the fractional part of the phase of *Div* in the same way that the accumulator in an accumulator-based ADPLL determines the fractional and integer parts of the phase of the *DCO*. Hence, the output of the accumulator in Fig. 2.34 is a digital word that represents a fractional number. The block labelled "Overflow Logic" in Fig. 2.34 is necessary to guarantee that the output of the accumulator is always a fractional number. In fact, the "Overflow Logic" block reads the output $out_{acc}[n]$ of the accumulator at the n-th edge of the reference oscillator. When $INC_{frac} + out_{acc}[n]$ is greater than 1, the output of $out_{ov}[n + 1]$ is equal to -1, otherwise $out_{ov}[n + 1]$ is zero. The architecture in Fig. 2.34 is also missing in the literature.

2.9 Comparison

In the next sections, we compare PFD-plus-TDC-based, TDC-based, flip-flop-based and accumulator-based ADPLLs in terms of phase-to-digital conversion, metastability and dynamic range of the TDC.

2.9.1 Phase-to-Digital Conversion

We recall that in Sect. 2.3, we classified the architectures of ADPLLs as "phase difference digitizing" or "phase digitizing". By comparing Figs. 2.15, 2.20 and 2.31, we deduce that PFD-plus-TDC-based, TDC-based an accumulator-based ADPLLs are described by equivalent phase-to-digital models when the integer part of the phase difference is zero. In fact, PFD-plus-TDC-based, TDC-based and accumulator-based ADPLLs can be classified as "phase difference digitizing" architectures when the integer part of the phase difference is zero.

By contrast, TDC-based ADPLLs with counters and accumulator-based ADPLLs can be considered as "phase digitizing" ADPLLs when the integer part of the phase difference is not zero, as can be deduced from the models in Figs. 2.19 and 2.30.

2.9.2 Metastability

Errors in the phase-to-digital conversion can be caused by metastability. Hence, it is important to compare the main ADPLL architectures in terms of the effects of metastability on the phase-to-digital conversion.

Metastability occurs in a flip-flop triggered by a clock when the instant at which the clock triggers the flip-flop is too close to the instant at which the input of the flip-flop changes logic level. Metastability can cause a D-flip-flop that is used as a sampler to produce sampling errors when the input and the clock are associated with two different clock domains. In PFD-plus-TDC-based, TDC-based and flip-flop-based ADPLLs, metastability can occur in the PFD, the TDC and the flip-flop. By contrast, in an accumulator-based ADPLL with retiming, metastability can occur in the TDC, the retiming block and the sampler at the output of the counter.

In a PFD with inputs *Ref* and *Div*, metastability can occur when the edges of *Ref* and *Div* are too close, namely when the phase difference is equal to zero. If the ADPLL is designed to have a non-zero phase difference at steady state, metastability issues in the PFD can be avoided. In a TDC, the effect of metastability on the output depends on the TDC architecture. In order to keep our analysis simple, we consider a delay-line-based TDC as an example architecture.

Figure 2.35 shows the block diagram of a delay-line-based TDC in which the phase-to-digital conversion is realized by propagating the start signal through a delay line and by sampling the outputs of the delays with D-flip-flops. Assume that all the delay elements in Fig. 2.35 are characterized by the same delay res. Hence, res is the *time resolution* of the TDC. Let us assume that N_{met} is the number of delay elements whose inputs transitions are close enough to the edge of $Stop$ to cause metastability. Hence, the maximum difference between the time interval that has to be measured and the TDC output is equal to N_{met} times res. The maximum difference between the ideal and the digitized phase-difference caused by metastability in PFD-plus-TDC-based,

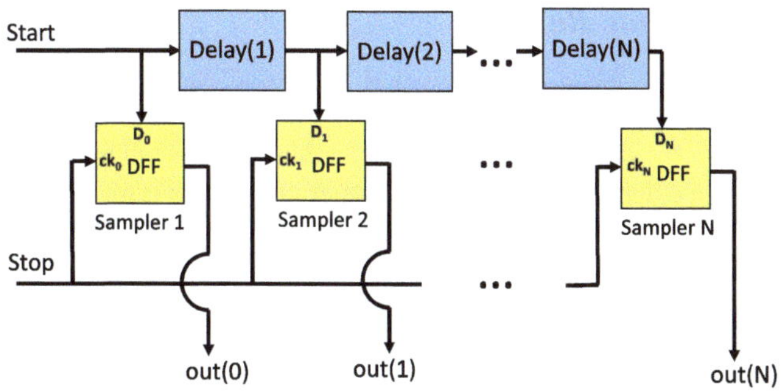

Fig. 2.35 Block diagram of a delay-line-based TDC

TDC-based and accumulator-based ADPLLs with a delay-line based TDC is equal to $res \cdot N_{met}/T_{Ref}$.

Metastability in the retiming flip-flop in Fig. 2.28 can produce a measurement of phase difference equal to minus one cycle of the reference clock when the actual phase difference is zero. In fact, one edge of Ref is always associated with one edge of the output Ref_R of the retiming flip-flop in Fig. 2.28, when the period of the DCO is much smaller than that of the reference oscillator. However, because of metastability in the retiming flip-flop, there can be an edge of *Ref* which is not associated with an edge of Ref_R. When there is one edge of *Ref* without a corresponding edge of Ref_R, an incorrect phase difference measurement equal to minus one cycle of the reference clock is produced.

Metastability in the block "D-flip-flop" at the output of the counter in Fig. 2.28 can produce a phase difference measurement included in the interval $[0, 2^b]$ DCO cycles, where b is the number of bits in the counter, when the actual phase difference is zero. In fact, we recall that the block "D-flip-flop" which samples the output of the counter in Fig. 2.28 represents an array of flip-flops that sample the counter output. Metastability in the flip-flops that sample the output of the counter in Fig. 2.28 can produce a difference in the interval $[0, 2^b]$ between the output word of the counter and that of the sampling flip-flops.

Assume that the output of the counter is associated with zero phase difference. Hence, a difference in the interval $[0, 2^b]$ between the sampled and the actual output word of the counter corresponds to a phase difference measurement included in the interval $[0, 2^b]$ DCO cycles when the actual phase difference is zero. Notice that $[0, 2^b]$ DCO cycles are approximately equal to $[0, 2^b/INC]$ cycles of the reference oscillator when the ADPLL has reached steady state. Therefore, metastability on the flip-flops that sample the counter output in an accumulator-based ADPLL has a much stronger impact on the difference between digitized and actual phase difference than metastability in the TDC in accumulator-based, PFD-plus-TDC and TDC-based ADPLLs.

2.9.3 TDC Dynamic Range

The dynamic range of a TDC is determined by the maximum duration of the time interval that the TDC can measure. The value of the dynamic range that is required by a TDC in an ADPLL depends on the ADPLL operation in terms of (i) maximum phase difference caused by noise and (ii) ratio between the frequencies of the reference oscillator and the DCO.

The maximum phase difference caused by noise in an ADPLL is related to the ADPLL noise transfer function, as shown in Chap. 4. In Chap. 4, we will show that the noise transfer functions of a TDC-based and an accumulator-based ADPLL are very similar when the integer part of the phase difference is zero. Assume that a TDC-based and an accumulator-based ADPLL are associated with the same noise transfer function. Hence, the maximum phase difference caused by noise is the same in the two ADPLL when the noise sources in the ADPLLs are the same and the integer part of the phase difference is zero. We conclude that TDC-based and accumulator-based ADPLLs require the same TDC dynamic range when they are associated with the same transfer function and the integer part of the phase difference is zero.

In Sect. 2.8, we have shown that a TDC-based and an accumulator-based ADPLL can include a sigma-delta modulator that determines N and INC, respectively, in order to synthesize a fractional ratio between the frequencies of the reference oscillator and the DCO. In these fractional-N TDC-based and accumulator-based ADPLLs, the maximum variations in N and INC, respectively, determine the required dynamic range of the TDC. The variations in N and INC are related to the output of the sigma-delta modulator. It is important to notice that the maximum variations in N and INC determine the variations in the phase difference with an integer part *different from zero*. The models that we have shown in Sects. 2.5.1 and 2.7.1, can be used to determine the maximum phase differences associated with the maximum variations in N and INC in a TDC-based and an accumulator-based ADPLL, respectively. The required TDC dynamic range is the same in a TDC-based and an accumulator-based ADPLL in which the maximum phase difference associated with the maximum variation in N and INC is the same.

2.10 Additional Design Considerations

2.10.1 Introducing Digital-Sigma-Delta-Modulators in the Architectures

A digital sigma-delta modulator is often introduced between the digital filter and the DCO in all of the ADPLL architectures that we have considered. A digital sigma-delta modulator between the digital filter and the DCO does not change how the phase-to-digital conversion is realized in the ADPLL architectures that wehave analysed.

A sigma-delta modulator followed by a DCO can be considered as an equivalent DCO with higher effective resolution.

TDCs with nth-order noise-shaping of the quantization error can be used in ADPLLs. An nth-order noise-shaping TDC is represented by an nth-order sigma-delta modulator in the phase-to-digital model. An nth-order noise-shaping TDC followed by a low pass filter can be considered as an equivalent TDC with higher effective resolution, as will be discussed in Chap. 3.

2.10.2 Selecting the Clock of the Digital Filter

The clock of the digital filter in an ADPLL has to guarantee that the digital filter samples and processes every measurement that is produced by the TDC.

When the integer part of the phase difference between *Ref* and *Div* or between *Ref* and Ref_R is equal to zero, the TDC in a TDC-based ADPLL or in an accumulator-based ADPLL produces one measurement of phase difference per cycle of the reference oscillator. Notice that only one cycle of *Div* or Ref_R is associated with a cycle of *Ref* when the integer integer part of the phase difference between *Ref* and *Div* or between *Ref* and Ref_R is equal to zero. Hence, in a TDC-based ADPLL, a digital filter that is clocked by *Ref* or *Div* samples and processes each measurement produced by the TDC when the integer part of the phase difference is zero. In an accumulator-based ADPLL, a digital filter clocked by *Ref* or Ref_R samples and processes each measurement produced by the TDC when the integer part of the phase difference is zero.

However, it is important to notice that the edges of *Div* and Ref_R are determined by those of the DCO. By contrast, the edges of *Ref* are determined by a crystal oscillator in ADPLLs that are used as frequency synthesizers. The noise associated with the edges of the reference signal is typically much lower than that in the DCO when the reference signal is provided by a crystal oscillator. A noisy clock can compromise the operation of the components in the digital filter. Hence, issues related to a noisy clock can be avoided by clocking the digital filter with the reference oscillator.

2.11 Conclusions

In this chapter, we have reviewed the operating principles of the main topologies of ADPLLs (PFD-plus-TDC-based, TDC-based and accumulator-based ADPLLs) in terms of the integer and fractional parts of the phase difference. We have derived models that describe the phase-to-digital conversion in each ADPLL architecture when the integer part of the phase difference is equal to or different from zero. We have discussed possible strategies to clock the digital filter in the various ADPLL architectures. We have also showed how to modify the ADPLL architectures to synthesize a fractional ratio between the frequencies of the reference oscillator and the DCO. Finally, we have compared the ADPLL architectures in terms of phase-to-digital conversion, TDC dynamic range, and metastability.

References

1. E. Temporiti, C. Weltin-Wu, D. Baldi, R. Tonietto, F. Svelto, "A 3 GHz Fractional All-Digital PLL With a 1.8 MHz Bandwidth Implementing Spur Reduction Techniques," *IEEE J. Solid-State Circuits*, vol. 44, no. 3, pp. 824–834, Mar. 2009.
2. C.-W. Yao, A. N. Willson, "A 2.8-3.2-GHz Fractional-N Digital PLL with ADC-Assisted TDC and Inductively Coupled Fine-Tuning DCO," *IEEE J. Solid-State Circuits*, vol. 48, no. 3, pp. 698–710, Mar. 2013.
3. V. Kratyuk, P. K. Hanumolu, U.-K. Moon, K. Mayaram, "A Design Procedure for All-Digital Phase-Locked Loops Based on a Charge-Pump Phase-Locked-Loop Analogy" *IEEE Trans. Circuits Sys. II: Express Briefs*, vol. 44, no. 3, pp. 247–251, Mar. 2007.
4. K. Nagaraj, A. S. Kamath, K. Subburaj, G Chattopadhyay, G. Nayak, S. S. Evani, N. P. Nayak, I. Prathapan, F. Zhang, B. Haroun, "Architectures and Circuit Techniques for Multi-Purpose Digital Phase Lock Loops", *IEEE Trans. Circuits Sys. I: Regular Papers*, vol. 60, no. 3, pp. 517–528, Apr. 2013.
5. C.-M. Hsu, M. Z. Straayer, M. H. Perrott, "A Low-Noise Wide-BW 3.6-GHz Digital $\Delta\Sigma$ Fractional-N Frequency Synthesizer With a Noise-Shaping Time-to-Digital Converter and Quantization Noise Cancellation," *IEEE J. Solid-State Circuits*, vol. 43, no. 12, pp. 2776–2786, Dec. 2008.
6. C.-C. Hung, S.-I. Liu, "A 40-GHz Fast-Locked All-Digital Phase-Locked Loop Using a Modified Bang-Bang Algorithm," *IEEE Trans. Circuits Sys. II: Express Briefs*, vol. 58, no. 6, pp. 321–325, Jun. 2011.
7. N. Da Dalt, "Theory and Implementation of Digital Bang-Bang Frequency Synthesizers for High Speed Serial Data Communications," *PhD Thesis*, RWTH Aachen University, Feb. 2007.
8. R. B. Staszewski, "State-of-the-Art and Future Directions of High-Performance All-Digital Frequency Synthesis in Nanometer CMOS," *IEEE Trans. Circuits Sys. I: Regular Papers*, vol. 58, no. 7, pp. 1497–1510, Jul. 2011.
9. R. B. Staszewski, P. T. Balsara, "All-Digital Frequency Synthesizer in Deep-Submicron CMOS," John Wiley & Sons Inc., 2006.
10. S. Mendel, C. Vogel, N. Da Dalt, "A Phase-Domain All-Digital Phase-Locked Loop Architecture Without Reference Clock Retiming," *IEEE Trans. Circuits Sys. II: Express Briefs*, vol. 56, no. 11, pp. 860–264, Nov. 2009.

Chapter 3
A Unifying Framework for TDC Architectures

3.1 Introduction

A TDC is an analog-to-digital converter that converts the duration t_{in} of a time interval to a digital word [1]. The duration t_{in} of the time interval that has to be measured is determined by the edges of two signals, *Start* and *Stop*, as illustrated in Fig. 3.1.

The analog-to-digital conversion is realized by counting how many *complete* reference time intervals are included in the time interval with unknown duration, as can be seen in Fig. 3.1. The reference time interval determines the time resolution of the TDC. The quantization error e_q is equal to the duration of the *fractional* reference time interval that is included between the edges of the *Start* and *Stop* signals.

In this chapter we summarize the operating principles of the main TDC architectures that comprise delay-line-based, oscillator-based, pulse-shrinking, Vernier-delay-line-based, Vernier-oscillator-based and GRO-based TDCs.

We show that all known architectures can be represented as equivalent single-scale or dual scale analog-to-digital converters.

3.2 Single Scale TDC and Flash Analog-to-Digital Converter

Time-to-Digital Converters that are based on a single scale can be related to flash ADCs.

A flash ADC [2] measures the analog quantity x (which is typically a voltage [2]) by comparing x with a series of ordered thresholds $th(k)$ with $k = 0, 1, ...$, as shown in Fig. 3.2. The difference $th(k) - th(k-1)$ is the voltage resolution [1] $res(k)$. In an ideal ADC, we assume that $res(k)$ is constant with respect to k and equal to the *nominal* voltage resolution res_n.

---

[1] In Time-to-Digital Converters x is the duration of a time interval. Therefore, in a TDC, the difference between two consecutive values of $th(k)$ is the *time* resolution.

F. Brandonisio and M. P. Kennedy, *Noise-Shaping All-Digital Phase-Locked Loops*, Analog Circuits and Signal Processing, DOI: 10.1007/978-3-319-03659-5_3,

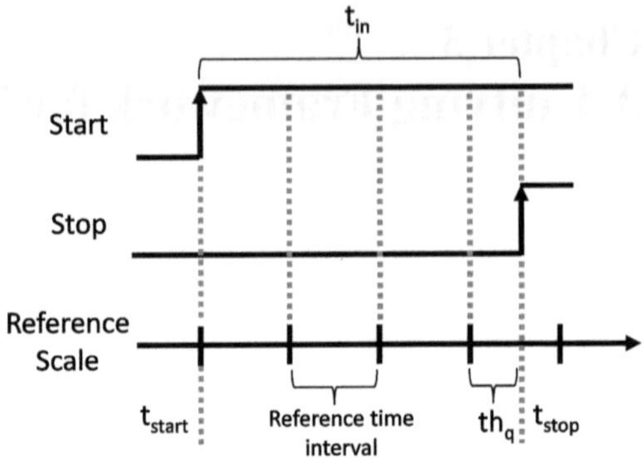

Fig. 3.1 Concept of a TDC. The TDC quantizes the duration of the time interval between the rising edges of the *Start* and *Stop* signals

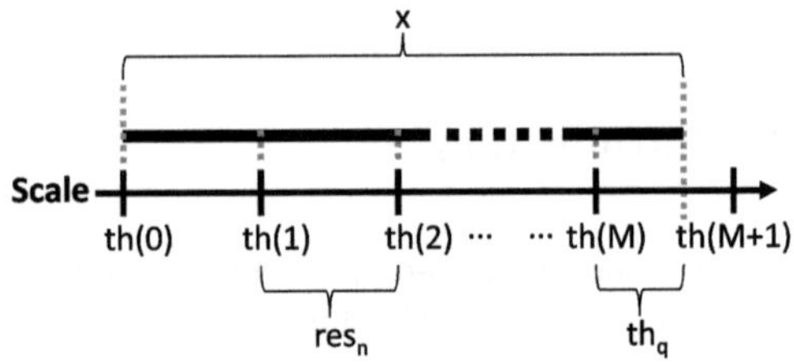

Fig. 3.2 Analog quantity x and reference scale in a flash Analog-to-Digital Converter

The operating principle of a flash ADC is to check *in parallel* whether or not x is greater than or equal to each threshold. If $th(k) \leq x < th(k+1)$, then the quantity x that has to be measured is given by:

$$x = k \cdot res_n + th_q, \tag{3.1}$$

where k is the value of the digital output word of the ADC, res_n is the nominal voltage resolution, and th_q is the quantization error associated with the analog to digital conversion.

Figure 3.3 shows the architecture of a flash ADC based on m comparators. Each comparator has two inputs A and B, and one output *out*. The output of each comparator in Fig. 3.3 is equal to binary one when A is greater than or equal to B, otherwise it is binary zero. If $th(k) \leq x < th(k+1)$, the outputs of each of the comparators associated with a value smaller than or equal to k is zero; otherwise it is one. The sequence of the outputs of the comparators is called a "thermometer code" [2]. A decoder produces a digital word out_{dec} equal to k from the outputs of the comparators, as can be seen in Fig. 3.3.

The output out_{dec} of an ideal flash ADC is given by:

$$out_{dec} = \text{floor}\left(\frac{x}{res_n}\right), \tag{3.2}$$

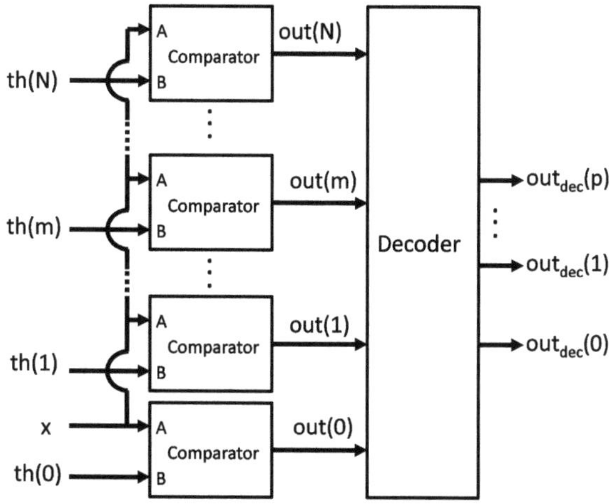

Fig. 3.3 Architecture of a flash ADC with comparators

where x and and res_n are the input and the nominal voltage resolution of the ADC, respectively, and floor$(\cdot)$ rounds to the largest integer that is smaller than or equal to the argument.

3.3 Notation

In this chapter, we will use the notation "(k)" in order to indicate quantities that are associated with the kth threshold is a TDC. When the thresholds $th(k)$ are determined by the rising edges of a digital signal, a discrete-time variable k can be associated with the instants at which these edges occur. Hence, the thresholds $th(k)$ can be seen as a discrete-time signal $th[k]$ when the thresholds $th(k)$ are determined by the rising edges of a digital signal. The square *brackets* in the notation "$[k]$" emphasize the discrete-time nature of the signal $th[k]$. In later chapters, we will study an ADPLL, which includes a TDC, as a discrete-time system and, therefore, we will use the discrete-time notation "$[k]$" for the ADPLL and TDC models.

3.4 The Delay-Line-Based TDC

The architecture of a delay-line-based TDC is shown in Fig. 3.4. The delay elements in Fig. 3.4 generate delayed versions of the signal *Start* with increasing delays. The D-flip-flops that are labelled "DFF" in Fig. 3.4 sample the delayed versions of the signal *Start* when the rising edge of the signal *Stop* occurs.

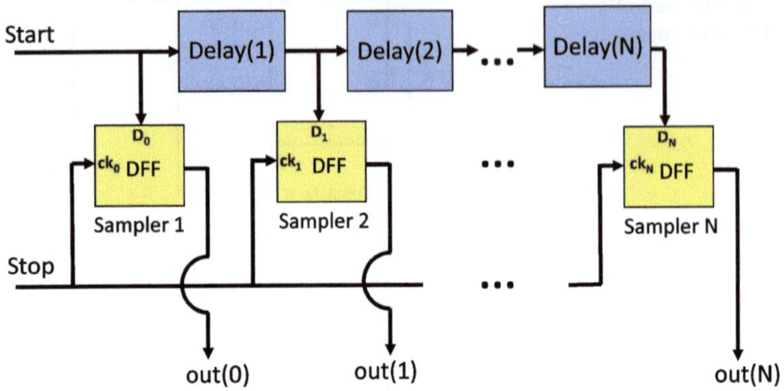

Fig. 3.4 Block diagram of a Delay-Line-based TDC implemented with D-Flip-Flops and Delay elements

Fig. 3.5 Timing diagram of the transitions of the D-Flip-Flops in a delay-line-based TDC

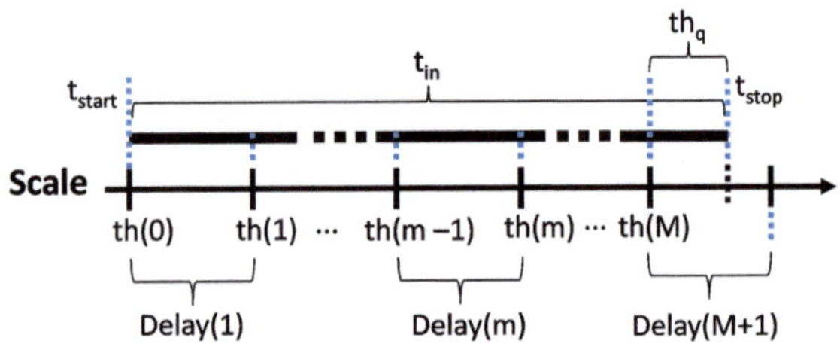

The operation of a delay-line-based TDC can be related to that of a flash ADC that is described in Sect. 3.2.[2] In fact, the transitions of the delay elements can be associated with a series of thresholds $th(m)$, where $th(m)$ is the sum of the delays of the first m delay elements of the delay line, as illustrated in Fig. 3.5. Moreover, the mth D-flip-flop operates as a comparator that produces a binary one at the output when the duration t_{in} of the time interval between the rising edges of *Start* and *Stop* is greater than or equal to $th(m)$; otherwise, it produces a binary zero.

The time resolution $res(m)$ of the delay line is the delay of the mth delay element, as can be seen by comparing Figs. 3.2 and 3.5.

In an ideal delay-line-based TDC, $res(m)$ is constant with respect to m and is equal to the nominal time resolution res_n. If the outputs of the Mth and $(M + 1)$th flip-flops of a delay-line-based TDC are zero and one, respectively, the duration of the time interval between the rising edges of *Start* and *Stop* is:

$$t_{in} = M \cdot res_n + th_q, \tag{3.3}$$

where res_n is the nominal time resolution and th_q is the quantization error, which is positive and smaller than res_n.

---

[2] A flash TDC performs comparisons with all quantization levels *simultaneously* in time. A TDC also compares to all levels but does it *sequentially* in time.

Table 3.1 Examples of delay-line-based TDCs with picosecond resolution

Work	Technology	Resolution [ps]
Staszewski [3]	90 nm	20
Yousif [4]	NEC Nb2.5 kA/cm	>1
Terabe [5]	0.13 μm	31
Lee [6]	90 nm	1.25
Lee [7]	90 nm	0.75
Wang [8]	FPGA	50

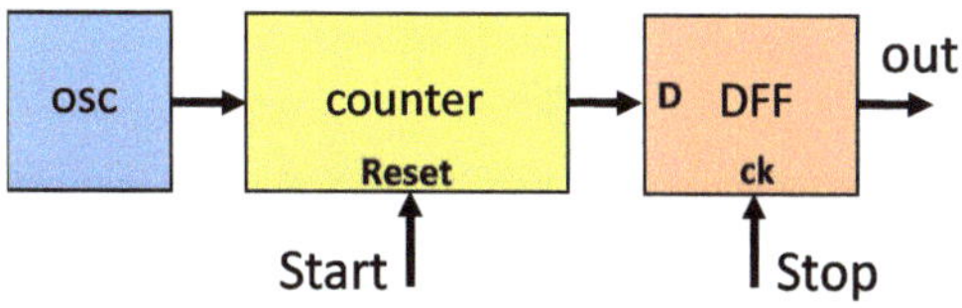

Fig. 3.6 Block diagram of an Oscillator-based TDC implemented with an oscillator and a counter

Table 3.1 summarizes some recent work on delay-line-based TDCs with picosecond resolution.

3.5 The Oscillator-Based TDC (Plus Delay Line TDC)

The simplest implementation of an oscillator-based TDC includes a free-running oscillator, a counter and a D-flip-flop, as shown in Fig. 3.6.

The counter is reset on the rising edges of *Start* and the D-flip-flop that is labelled "DFF" in Fig. 3.6 samples the output of the counter on the rising edges of *Stop*. The counter counts how many complete periods of the oscillator are included between the rising edges of *Start* and *Stop*.

Similar to a delay-line-based TDC, the operating principle of the oscillator-based TDC can be related to that of the flash ADC described in Sect. 3.2. In fact, a series of thresholds th_m can be associated with the rising edges of the oscillator, as shown in Fig. 3.7. $th(m)$ is equal to the sum of the first m periods of the oscillator. Moreover, the system "oscillator plus counter" operates as a comparator in a flash TDC. In fact, the "oscillator plus counter" increments the output of the counter by one when t_{in} is greater than $th(m)$; otherwise, the output of the counter is constant.

The duration of the mth period of the oscillator is the time resolution $res(m)$ of the oscillator-based TDC, as can be seen by comparing Figs. 3.7 and 3.2.

In an ideal oscillator-based TDC, the duration of the periods of the oscillator is constant and is equal to the nominal time resolution res_n.

Assume that the output of the counter of an ideal oscillator-based TDC is equal to M when the output of the counter is sampled. Hence, the duration of the time interval between the rising edges of *Start* and *Stop* is:

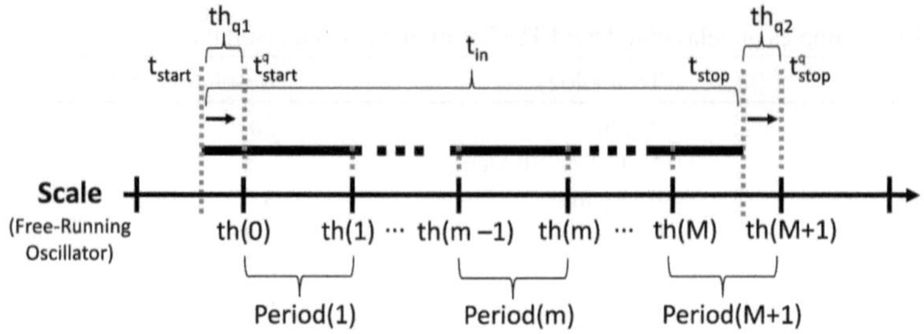

Fig. 3.7 Timing diagram of the edges of the oscillator in an Oscillator-based TDC

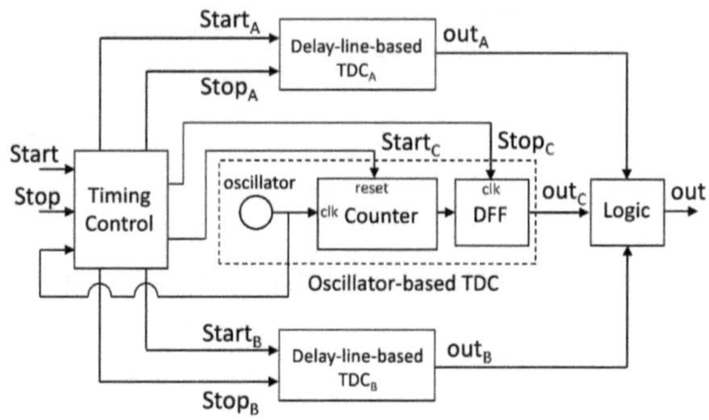

Fig. 3.8 Block diagram of an oscillator-based TDC combined with two delay-line-based TDC that provide fine resolution

$$t_{in} = M \cdot res_n + th_{q1} - th_{q2}, \tag{3.4}$$

where res_n is the nominal time resolution, and th_{q1} and th_{q2} are the durations of the time intervals associated with the quantization errors in Fig. 3.7. By definition, th_{q1} and th_{q2} are positive and smaller than res_n.

The duration of the maximum time interval that an oscillator-based TDC can measure is proportional to 2^b, where b is the number of bits of the counter. Hence, an oscillator-based TDC with a small number of components can measure the duration of a time interval that is much larger than its time resolution. In order to exploit this feature, an oscillator-based TDC can be combined with other TDC architectures, as illustrated in Fig. 3.8.

In fact, an oscillator-based TDC can be used to measure the duration of a time interval with a *coarse* time resolution. Then, another TDC architecture with a *fine* time resolution has to measure only the duration of the time intervals th_{q1} and th_{q2} that are always smaller than a period of the free-running oscillator in the oscillator-based TDC, as can be seen in Fig. 3.9.

Table 3.2 illustrates some recent publications on TDCs in which an oscillator-based TDC and another TDC provide measurements with coarse and fine resolutions, respectively; the resolution reported in Table 3.2 is the *fine* time resolution.

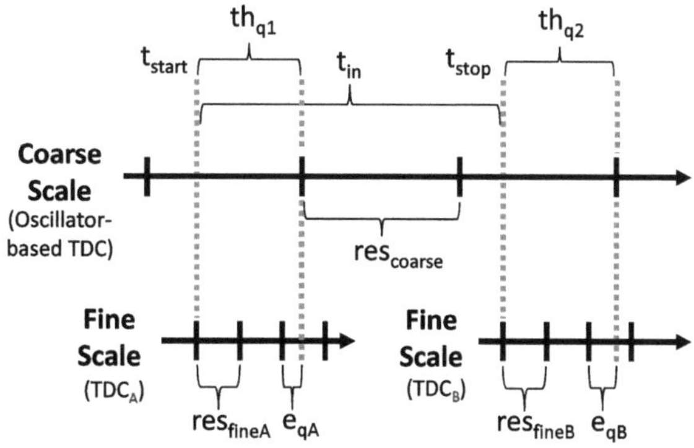

Fig. 3.9 Timing diagram of an oscillator-based TDC combined with two delay-line-based TDC that provide fine resolution

Table 3.2 Examples of TDCs with picosecond resolution that combine oscillator-based TDCs with other TDC architectures

Work	Technology	Resolution [ps]
Chen [9]	0.35 μm	50
Song [10]	FPGA	45
Jansson [11]	0.35 μm	12
Nissinen [12]	0.13 μm	13
Jansson [13]	0.35 μm	9.6

3.6 Dual Scale Time-to-Digital Converters and the Vernier Method

An analog-to-digital converter can be realized by means of two scales. In some TDC architectures, it is possible to associate two time scales with two delay lines or two oscillators, such as in Vernier TDCs. In other architectures, such as pulse shrinking TDCs and GRO-based TDCs, the two scales are not evident from the architectures of the circuit. However, these TDCs can be represented with models that include two scales.

3.6.1 *The Vernier Method*

The operation of a Vernier TDC is based on that of the Vernier Caliper, invented by Pierre Vernier in 1631 [14]. In the Vernier method , the quantity x (which is a physical length in a Vernier Caliper) is measured by means of two scales A and B that are defined by two series of thresholds $th_A(m)$ and $th_B(k)$ with $m = 0, 1, ...$ and $k = 0, 1, ...$ respectively. Figure 3.10 illustrates the scales A and B and the input x. The differences $th_A(m) - th_A(m-1)$ and $th_B(k) - th_B(k-1)$ are the length[3] resolutions $res_A(m)$ and $res_B(k)$, respectively. In an *ideal* Vernier Caliper,

[3] In a Vernier TDC, $res_A(m)$ and $res_B(k)$ are the *time* resolutions.

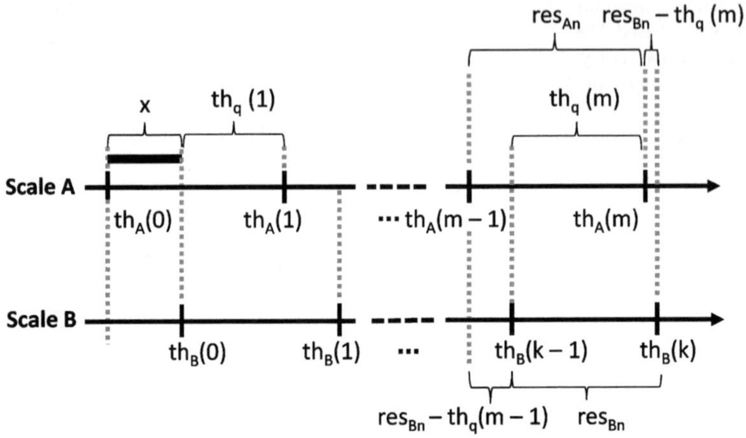

Fig. 3.10 Analog quantity x and reference scales in a Vernier Caliper when one threshold th_A

we assume that $res_A(m)$ and $res_B(k)$ are constant with respect to m and k, and equal to the nominal metric[4] resolutions res_{An} and res_{Bn}, respectively.

By means of the Vernier method, a TDC can theoretically achieve a nominal time resolution res_{Cn} that is given by:

$$res_{Cn} = res_{An} - res_{Bn}. \quad (3.5)$$

In order to illustrate how the Vernier method works, we assume that $x > res_{Cn}$ and $res_{An} > res_{Bn} > res_{An}/2$. When $res_{An} > res_{Bn} > res_{An}/2$, we have that the number of thresholds th_B that are included in the interval $[th_A(m-1), th_A(m)]$ can only be (i) one or (ii) two. When $x > res_{Cn}$, the interval $[th_A(0), th_A(1)]$ includes only one value of th_B, as shown in Fig. 3.10. Moreover, assume that case (i) applies to all the intervals $[th_A(m-1), th_A(m)]$ with $m = 1, 2, ..., M$. By contrast, case (ii) applies to the interval $[th_A(M), th_A(M+1)]$, as illustrated in Fig. 3.11.

Let us define $th_q(m)$ as the length of the interval between $th_A(m)$ and the largest value of $th_B(k)$ that is smaller than $th_A(m)$. $th_q(m)$ is given by different equations for cases (i) and (ii). For case (i), from Fig. 3.10 we can write:

$$th_q(m) = th_q(m-1) + res_{Cn}. \quad (3.6)$$

For case (ii), from Fig. 3.11, we can write:

$$th_q(M+1) = th_q(M) + res_{Cn} - res_{Bn}. \quad (3.7)$$

Moreover, from Fig. 3.10, we have:

[4] In a Vernier TDC, res_{An} and res_{Bn} are the nominal *time* resolutions.

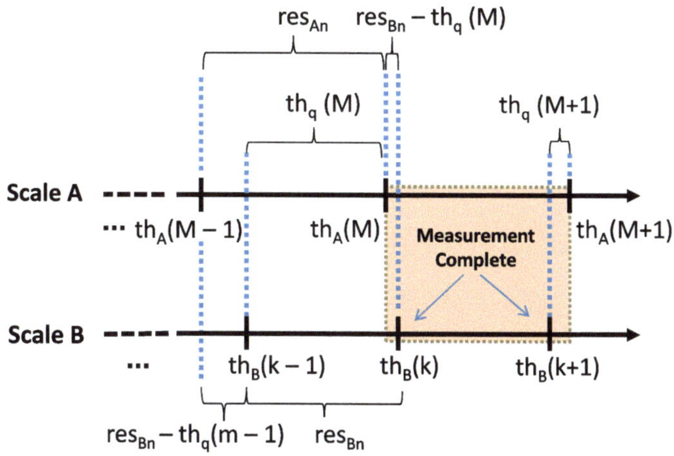

Fig. 3.11 Analog quantity x and reference scales in a Vernier Caliper when the measurement is complete

$$th_q(1) = res_{An} - x. \tag{3.8}$$

From Eqs. (3.6) and (3.8), we have:

$$th_q(m) = res_{An} - x + (m-1)\left(res_{An} - res_{Bn}\right). \tag{3.9}$$

Assume that at least one threshold th_B is included in the interval $[th_A(m-1), th_A(m)]$. The length of the interval between $th_A(m)$ and the threshold th_B that is included in $[th_A(m-1), th_A(m)]$ is $res_{Cn} + th_q(m-1))$, as can be seen from Fig. 3.10. When $res_{Cn} + th_q(m-1))$ is greater than or equal to res_{Bn}, there is a second threshold th_B in $[th_A(m-1), th_A(m)]$. Thus, the condition for case (ii) is:

$$res_{Cn} + th_q\,(M) \geq res_{Bn}. \tag{3.10}$$

By assuming that $m = M$ and substituting Eq. (3.9) into Eq. (3.10), we can write:

$$x - M \cdot res_{Cn} \leq res_{Cn}. \tag{3.11}$$

Moreover, notice that $th_q(m)$ is smaller than res_{An} for any m, by definition. Thus, from Eq. (3.9), with $m = M + 1$, we can write:

$$x - M \cdot res_{Cn} \geq 0. \tag{3.12}$$

From Eqs. (3.11) and (3.12), we have:

$$0 \leq x - M \cdot res_{Cn} \leq res_{Cn}. \tag{3.13}$$

From Eq. (3.13), we can write:

$$x = M \cdot res_{Cn} + th_{err}, \tag{3.14}$$

where $0 \leq th_{err} \leq res_{Cn}$.
Finally, let us relate th_{err} to $th_q(M)$. From Eq. (3.14) we can write:

$$th_{err} = x - M \cdot res_{Cn}. \tag{3.15}$$

From Eq. (3.9), with $m = M + 1$, we can write:

$$x - M \cdot res_{Cn} = res_{An} - th_q\,(M + 1)\,. \tag{3.16}$$

From Eqs. (3.15) and (3.16), we have that

$$th_{err} = res_{An} - th_q\,(M + 1)\,. \tag{3.17}$$

Hence, by multiplying res_{Cn} by the number M of consecutive pairs of thresholds th_A that include only one threshold th_B, we have a measure of x with an error given by Eq. (3.17), which is smaller than res_{Cn}. We conclude that $M \cdot res_{Cn}$ is a measure of x with resolution res_{Cn} .

3.6.2 The Vernier Method as a Case of Sigma-Delta Modulation

The Vernier method that we described in Sect. 3.6.1 can be related to the model of a first order sigma-delta modulator. As mentioned in Sect. 3.6.1, and as can be seen from Figs. 3.10 and 3.11, the number of thresholds th_B that are included in the interval $[th_A(m-1), th_A(m)]$ can be one or two. The distance between $th_A(m)$ and the threshold th_b that is included in $[th_A(m-1), th_A(m)]$ is equal to $res_{Cn}+th_q(m-1)$, as can be seen from Fig. 3.10. Hence, when $res_{Cn}+th_q(m-1)$ is smaller than res_{Bn}, there is only one threshold th_B included in the interval $[th_A(m-1), th_A(m)]$, and Eq. (3.6) applies. By contrast, when $res_{Cn} + th_q(m-1)$ is greater than or equal to res_{Bn}, there are two thresholds th_B included in the interval $[th_A(m-1), th_A(m)]$, and we can write:

$$th_q(m) = res_{Cn} - res_{Bn} + th_q(m-1). \tag{3.18}$$

Notice that we can obtain Eq. (3.7) from Eq. (3.18) by assuming that $m = M + 1$. An equation that gives $th_q(m)$ for any interval $[th_A(m-1), th_A(m)]$ is:

$$th_q(m) = res_{Cn} + th_q(m-1) - res_{Bn} \cdot \text{floor}\left(\frac{res_{Cn} + th_q(m-1)}{res_{Bn}}\right), \tag{3.19}$$

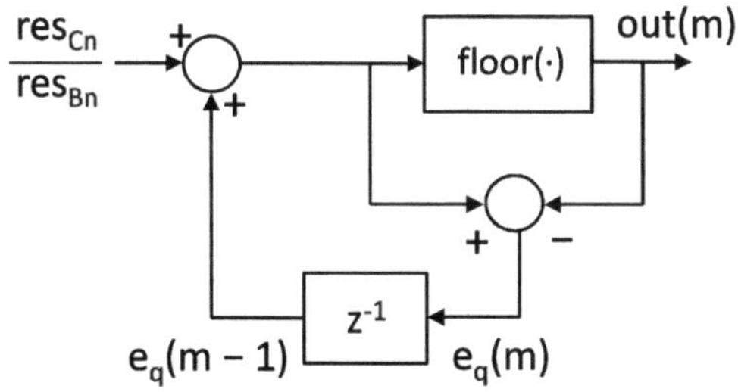

Fig. 3.12 Model of a sigma-delta modulator associated with the position of the thresholds of the scales of the Vernier method

where the floor(·) operator rounds to the largest integer that is smaller than or equal to the argument. The floor(·) operator in Eq. (3.19) is equal to zero when $res_{Cn} + th_q(m-1)$ is smaller than res_{Bn}, and Eq. (3.19) is equivalent to Eq. (3.6). By contrast, when $res_{Cn} + th_q(m-1)$ is greater than or equal to $\geq res_{Bn}$, the floor(·) operator in Eq. (3.19) takes the value one, making Eq. (3.19) equal to Eq. (3.18). From Eq. (3.19), we can write:

$$e_q(m) = \frac{res_{Cn}}{res_{Bn}} + e_q(m-1) - \text{floor}\left(\frac{res_{Cn}}{res_{Bn}} + e_q(m-1)\right), \tag{3.20}$$

where $e_q(m) = th_q(m)/res_{Bn}$. Finally, Eq. (3.20) for $m = 1, 2, \ldots$ can be associated with the first-order sigma-delta modulator in Fig. 3.12. The quantization error of the sigma-delta modulator in Fig. 3.12 is equal to $e_q(m)$, and is given by Eq. (3.20).

The output $out(m)$ of the first-order sigma-delta modulator in Fig. 3.12 is zero or one when the number of thresholds th_B in the interval $[th_A(m-1), th_A(m)]$ is one or two, respectively. Note that the quantity x that has to be measured is associated with the initial value $e_q(0)$ of the quantization error of the sigma-delta-modulator in Fig. 3.12 that is defined by:

$$e_q(0) = \frac{res_{An} - x}{res_{Bn}}. \tag{3.21}$$

3.7 The Vernier Delay-Line-Based TDC

The Vernier method described in Sect. 3.6.1 can be used to implement a Vernier Delay-Line-based TDC. The architecture of a Vernier Delay-Line-based TDC features two delay lines *A* and *B* with delays $Delay_A(m)$ and $Delay_B(k)$ respectively, as shown in Fig. 3.13. The flip-flops sample the delayed versions of *Start* when the positive-going edges of the delayed versions of *Stop* occur.

The operation of a Vernier delay-line-based TDC can be related to that of the Vernier Caliper that is described in Sect. 3.6.1. In fact, the transitions of the delay elements can be associated with a series of thresholds $th_A(m)$ and $th_B(k)$. $th_A(m)$ and $th_B(k)$ are the sums of the delays of the first *m* delay elements of the delay lines *A* and *B*, respectively, as shown in Fig. 3.14.

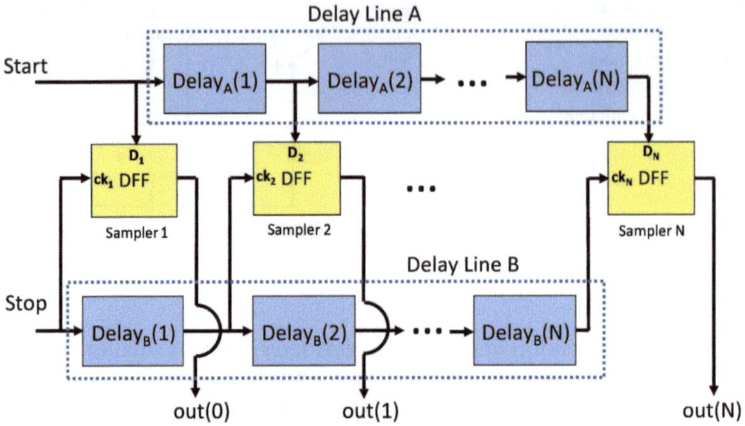

Fig. 3.13 Block diagram of a Vernier Delay-Line-based TDC implemented with two delay lines

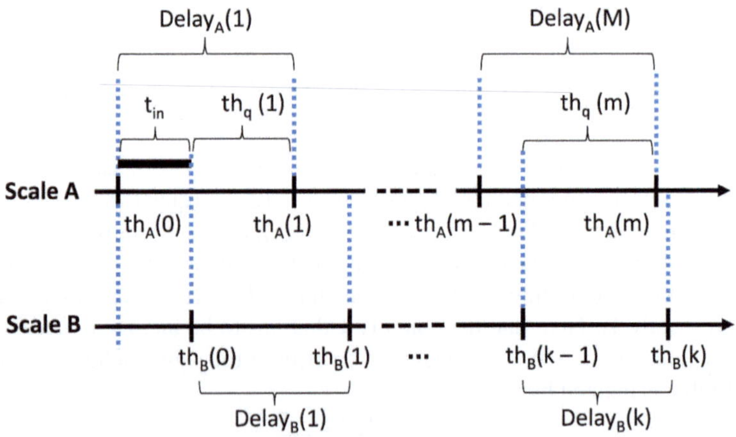

Fig. 3.14 Timing diagram of the transitions of the delay elements in a Vernier Delay-Line-based TDC

The time resolutions $res_A(m)$ and $res_B(k)$ of the delay lines of a Vernier delay-line-based TDC are equal to $Delay_A(m)$ and $Delay_B(k)$, as can be seen by comparing Figs. 3.14 and 3.10. In an *ideal* Vernier delay-line-based TDC, we assume that $res_A(m)$ and $res_B(k)$ are constant with respect to m and k, and equal to the *nominal* time resolutions res_{An} and res_{Bn}, respectively.

As mentioned in Sect. 3.6.1, we can have one or two thresholds th_B in the interval $[th_A(m-1), th_A(m)]$ when $res_{An} > res_{Bn} > res_{An}/2$. A measure of t_{in} is the number of consecutive intervals $[th_A(m-1), th_A(m)]$ that include one threshold th_B, starting from $m = 1$. The output signals of the flip-flops in Fig. 3.13 depend on the number of thresholds th_B that are included in the intervals $[th_A(m-1), th_A(m)]$. In fact, $th_A(m)$ and $th_B(m)$ are the instants of the rising edges of the input and the clock of the mth flip-flop in Fig. 3.13, respectively. When there is one threshold th_B

Table 3.3 Examples of Vernier delay-line based TDCs with picosecond resolution

Work	Technology	Resolution [ps]
Nose [15]	90 nm	1
Henzler [16]	90 nm	4.7
Temporiti [17]	65 nm	$\simeq$8
Amiri [18]	FPGA	75
Xing [19]	Simulation	22
Rashidzadeh [20]	FPGA	75
Vercesi [21]	65 nm	4.8

in the intervals $[th_A(m-1), th_A(m)]$ with $m = 1, 2, ..., M$, the rising edges of the inputs of the first M flip-flops occur *before* those of their clocks. Thus, the output signals of the first M flip-flops are binary zero when there is one threshold th_B in the intervals $[th_A(m-1), th_A(m)]$, with $m = 1, 2, ..., M$. By contrast, the output signals of the remaining flip-flops are equal to binary one.

In an *ideal* Vernier delay-line-based TDC, we assume that $res_A(m)$ and $res_B(k)$ are constant with respect to m and k, and equal to the *nominal* time resolutions res_{An} and res_{Bn}, respectively. If the output signals of the first M flip-flops of an ideal Vernier delay-line-based TDC are binary zero, the duration of the time interval between the rising edges of *Start* and *Stop* is given by Eq. (3.14), where $res_{Cn} = res_{An} - res_{Bn}$ and th_{err} is defined by (3.17).

Table 3.3 shows some recent works on Vernier delay-line based TDCs with picosecond resolution.

3.8 The Vernier Oscillator-Based TDC

The architecture of a Vernier oscillator-based TDC is shown in Fig. 3.15.

A Vernier oscillator-based TDC comprises two oscillators A and B with periods $Period_A(m)$ and $Period_B(k)$ respectively, as shown in Fig. 3.16. The oscillators A and B are enabled after the rising edges of *Start* and *Stop* occur, respectively. The oscillators are disabled when the phase-frequency detector in Fig. 3.15 detects that two rising edges of the oscillator B are included between two consecutive rising edges of the Oscillator A. We will explain the operation of the phase-frequency detector later in this section.

The concept of a Vernier oscillator-based TDC is the same as the Vernier caliper that is explained in Sect. 3.6.1. In fact, the rising edges of the oscillators can be associated with a series of thresholds $th_A(m)$ and $th_B(k)$. $th_A(m)$ and $th_B(k)$ are the sums of the durations of the first m periods of the oscillators A and B, respectively, as can be seen in Fig. 3.16. The periods $period_A(m)$ and $period_B(k)$ of the oscillators are the time resolutions $res_A(m)$ and $res_B(m)$, respectively, as can be seen by comparing Figs. 3.16 and 3.10.

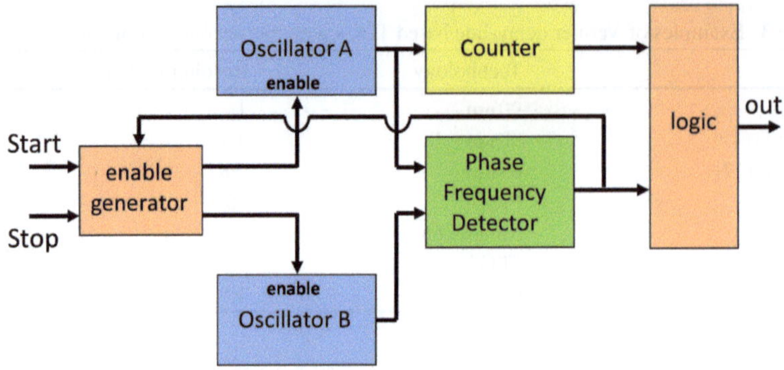

Fig. 3.15 Block diagram of a Vernier oscillator-based TDC implemented with two oscillators

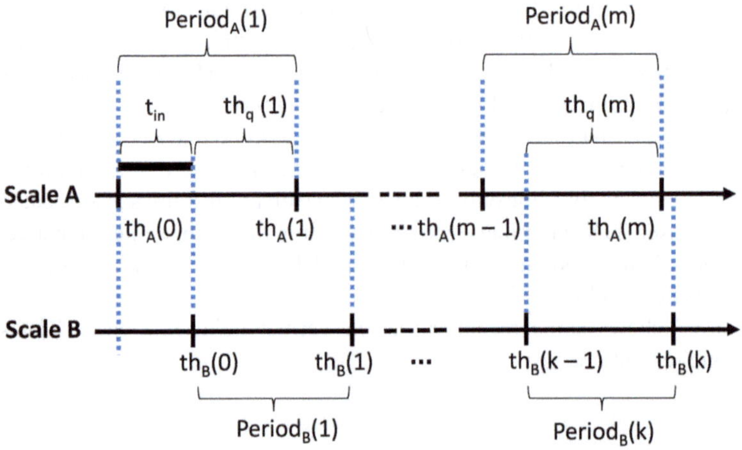

Fig. 3.16 Timing diagram of the edges of the oscillators in a Vernier Oscillator based TDC

In an *ideal* Vernier oscillator-based TDC, the periods of the oscillators A and B are constant and equal to the *nominal* time resolutions res_{An} and res_{Bn}, respectively. In an ideal Vernier oscillator-based TDC, there are only one or two thresholds th_B in the interval $[th_A(m-1), th_A(m)]$ when $res_{An} > res_{Bn} > res_{An}/2$, as we have shown for the Vernier caliper in Sect. 3.6.1. Similar to a Vernier delay-line-based TDC, a Vernier oscillator-based TDC produces a measure of the duration of the time interval t_{in} between the rising edges of *Start* and *Stop* by counting how many consecutive intervals of the forms $[th_A(m-1), th_A(m)]$ include one threshold th_B, starting from $m = 1$. The phase-frequency detector in an ideal Vernier-oscillator based TDC detects the first interval $[th_A(m-1), th_A(m)]$ with two thresholds th_B, starting from $m = 1$. In fact, the phase-frequency detector measures a positive phase difference between the oscillators A and B for the first consecutive intervals $[th_A(m-1), th_A(m)]$ that include one threshold th_B, as shown in Fig. 3.17.

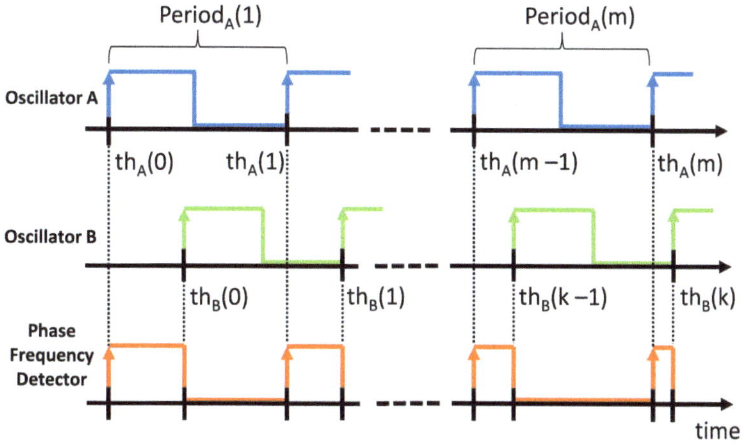

Fig. 3.17 Timing diagram of the output signals of the oscillators and of the phase-frequency detector when *one* edge of the Oscillator B is included between two consecutive edges of the Oscillator A

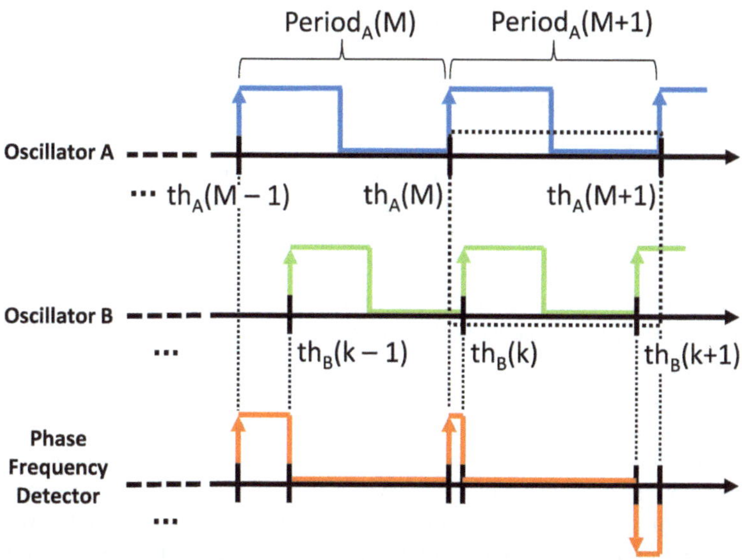

Fig. 3.18 Timing diagram of the output signals of the oscillators and of the phase-frequency detector when *two* edges of the Oscillator B are included between two consecutive edges of the Oscillator A

By contrast, the phase-frequency detector measures a *negative* phase difference between the oscillators A and B for the first interval $[th_A(m-1), th_A(m)]$ that includes *two* thresholds th_B, as shown in Fig. 3.18.

Assume that the phase difference that is measured by the phase-frequency detector is negative for the first time in the interval $[th_A(M), th_A(M+1)]$, as illustrated in

Table 3.4 Vernier Oscillator-based TDCs with picosecond resolution in the literature

Work	Technology	Resolution [ps]
Chen [22]	0.35 μm	37
Mollah [23]	Simulation only	10
Junnarkar [24]	FPGA	41
Yu [25]	0.13 μm	8
Chen [26]	FPGA	1.58

Fig. 3.18. The oscillations of the oscillators are disabled at $th_A(M+1)$ and the output of the counter is $M+1$. Thus, the duration of the time interval between the rising edges of *Start* and *Stop* is given by Eq. (3.14) where $res_{Cn} = res_{An} - res_{Bn}$ and th_{err} is given by (3.17).

Table 3.4 summarizes some recent publications on Vernier Oscillator-based TDCs with picosecond resolution.

3.9 The Pulse-Shrinking TDC

A pulse-shrinking TDC illustrated in Fig. 3.19 is based on a pulse generator, a pulse-shrinking delay line and a counter.

The pulse generator generates a digital pulse which is equal to a binary one during the time interval that has to be measured, and a binary zero otherwise. The pulse-shrinking delay line delays the input digital pulse by a specific time interval and reduces its duration by $\Delta Pulse(m)$. Control over the shrinking mechanism is obtained by introducing controlled non-uniformities in the delay line, namely by designing some delay elements with a delay that is different from that of the other delay elements.

Because the output of the delay line is fed back to its input, a single pulse from the pulse generator circulates in the loop of the delay line a number of times. A cycle is completed each time the pulse completes a loop. The duration of a cycle is determined by the total delay of the delay line. For every cycle, a pulse is produced at the output of the delay line, as shown in Fig. 3.20. The time interval between two edges of consecutive pulses at the output of the delay line is equal to the duration of a cycle. The duration of the pulse circulating in the loop decreases by the quantity $\Delta Pulse(m)$ every cycle. Hence, the durations of pulses that are produced at the output of the delay line also reduce every cycle, as can be seen in Fig. 3.20.

Assume that the $(M+1)$th cycle is the last one associated with a pulse at the output of the delay line, as shown in Fig. 3.21. Note that the duration of the pulse in Fig. 3.21 has to be smaller than $\Delta Pulse(m+1)$. The output of the "pulse extinction detector" is a binary one when the duration of the mth pulse is smaller than $\Delta Pulse(m+1)$; otherwise, it is a binary zero. The output of the counter in Fig. 3.19 is equal to the number of pulses that are produced by the delay line.

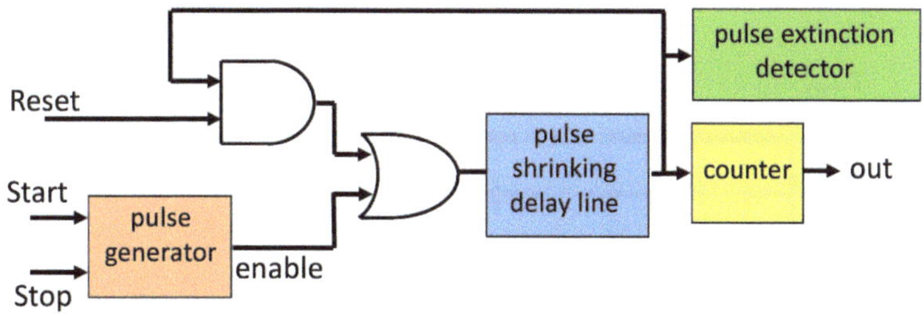

Fig. 3.19 Block diagram of a Pulse-Shrinking TDC implemented with a delay line

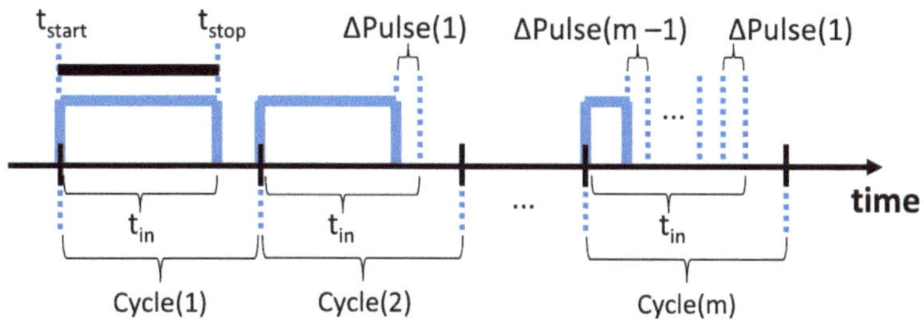

Fig. 3.20 Pulses at the output of the delay line in a Pulse-Shrinking TDC

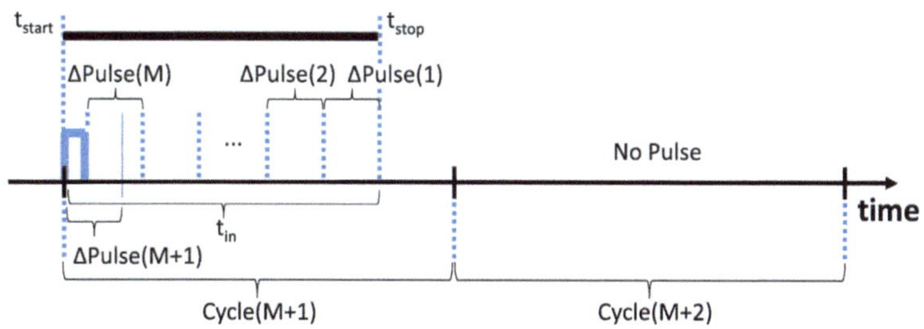

Fig. 3.21 Last pulse at the output of the delay line in a Pulse-Shrinking TDC

By redrawing the timing diagram of the $(M+1)$th cycle that is shown in Fig. 3.21, we can obtain a timing diagram of the variations in the durations of the pulses at the output of the delay line, as illustrated in Fig. 3.22.

The operating principle of a pulse-shrinking TDC is similar to that of the flash ADC described in Sect. 3.2. The series of thresholds $th(m)$ in Fig. 3.7 is associated with the distances between the rising and falling edges of the pulses. In particular, $th(m)$ is equal to the sum of the first m values of $\Delta Pulse$, namely the difference between the durations of the pulses associated with the first and the $(m+1)$th cycles. The "pulse extinction detector" in Fig. 3.19 operates like a comparator in a flash TDC. In fact, the output of the "pulse extinction detector" is a binary one when th_m is greater than t_{in}, otherwise it is a binary zero. The value $\Delta Pulse(m)$ is the time resolution $res(m)$ of a pulse-shrinking TDC, as can be seen by comparing Figs. 3.22 and 3.2.

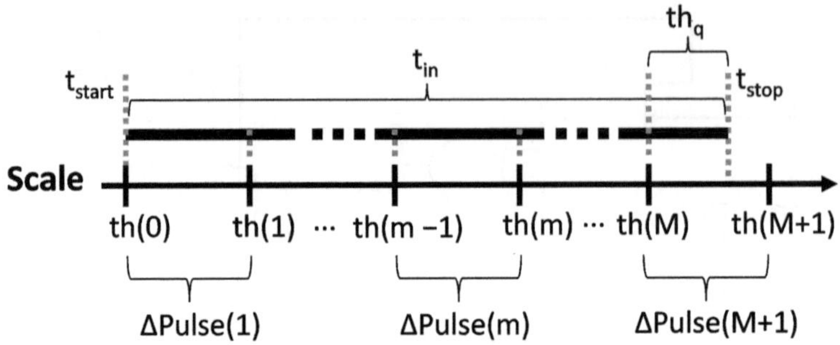

Fig. 3.22 Diagram of the durations of the pulses at the output of the delay line in a Pulse-Shrinking TDC

Table 3.5 Pulse-shrinking TDCs with picosecond resolution in the literature

Work	Technology	Resolution [ps]
Chen [27]	0.35 μm	53
Karadamoglou [29]	0.35 μm	50
Chen [28]	0.8 μm	68
Liu [30]	0.13 μm	20
Szplet [31]	FPGA	42

In an *ideal* pulse-shrinking TDC, we assume that the values of $\Delta Pulse(m)$ are constant and equal to the *nominal* time resolution res_n of the pulse-shrinking TDC. Assume that the output of the counter in an ideal pulse-shrinking TDC is equal to $(M + 1)$ when the output of the "pulse extinction detector" switches from zero to one, namely when the last pulse occurs. Thus, the duration of the initial digital pulse is given by Eq. (3.3) with res_n equal to the nominal time resolution of the pulse-shrinking TDC, and th_q equal to the quantization error. th_q is the duration of the last digital pulse in this case.

Table 3.5 summarizes some recent work on pulse-shrinking TDCs with picosecond resolution. Notice that in some publications, such as [27], the pulse-shrinking TDC is referred to as a "Cyclic" TDC.

3.10 The Gated-Ring-Oscillator-Based TDC

The architecture of a GRO-based TDC shown in Fig. 3.23 can be obtained from that of the Oscillator-based TDC in Fig. 3.6 by replacing the oscillator with a Gated-Ring-Oscillator. The GRO is enabled for the duration t_{in} of the time interval that has to be measured. A counter counts how many complete periods of the GRO are included in the time interval that has to be measured. This operation is the same as that of an oscillator-based TDC. However, the oscillator in an oscillator-based TDC is free-running. As a consequence, the main difference between Oscillator- and GRO-based TDCs is the state of the oscillator at the beginning of every measurement. In an

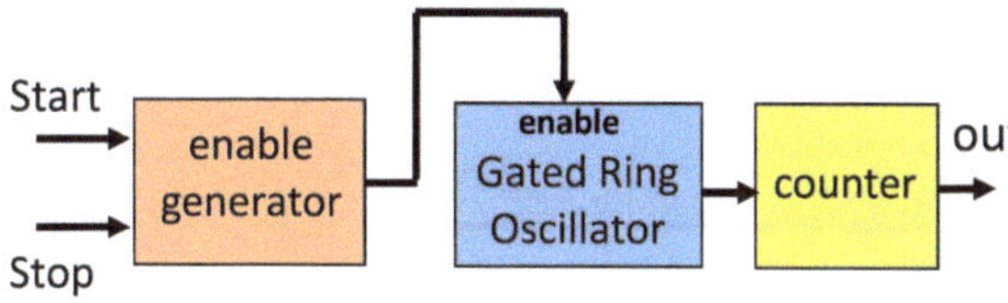

Fig. 3.23 Block diagram of an GRO-based TDC implemented with a Gated-Ring-Oscillator and a counter

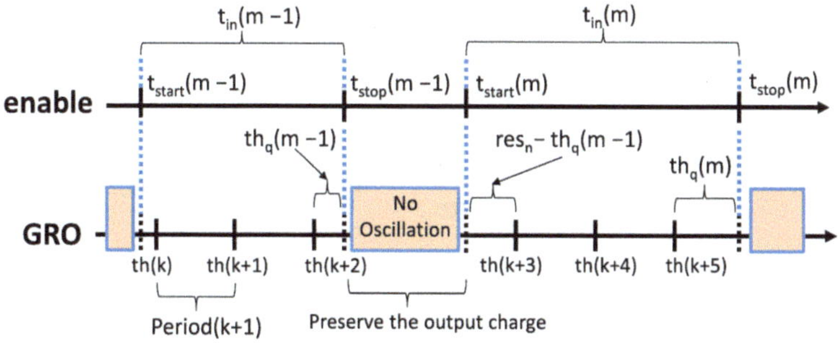

Fig. 3.24 Timing diagram of the rising edges of the GRO and of the transitions of the enable signal in a GRO-based TDC

oscillator-based TDC, the state of the oscillator at the beginning of a measurement is always *zero*. By contrast, at the beginning of the current measurement, the state of the GRO is the same as that at the end of the previous measurement. The state of the GRO between consecutive measurements is usually preserved by storing the output charge of the GRO. Thus, consecutive measurements are independent and dependent in Oscillator- and GRO-based TDCs, respectively.

Figure 3.24 shows the timing diagram of the rising edges of the GRO and of the transitions of the enable signal in a GRO-TDC.

The operating principle of a GRO-based TDC can be related to that of the flash ADC that is described in Sect. 3.2, when considering a single measurement. The time resolution $res(m)$ is the duration of the mth period of the GRO. However, a GRO-based TDC can be associated with a first-order sigma-delta modulator in order to model the dependence between consecutive measurements, as demonstrated hereafter. Assume an ideal GRO-based TDC with the durations of the complete periods of the GRO being constant and equal to the nominal time resolution res_n.

The timing diagram in Fig. 3.25 is equivalent to that in Fig. 3.24 in order to predict the output of an ideal GRO-based TDC at the end of every measurement. The output $out(m)$ of the counter at the end of the mth measurement is equal to the number of complete periods of the GRO in the time interval $[t_{start}(m), t_{start}(m+1)]$. From Fig. 3.25, we have that $out(m)$ is given by:

$$out(m) = \text{floor}\left(\frac{t_{in}(m) - res_n + th_q\,(m-1)}{res_n}\right), \tag{3.22}$$

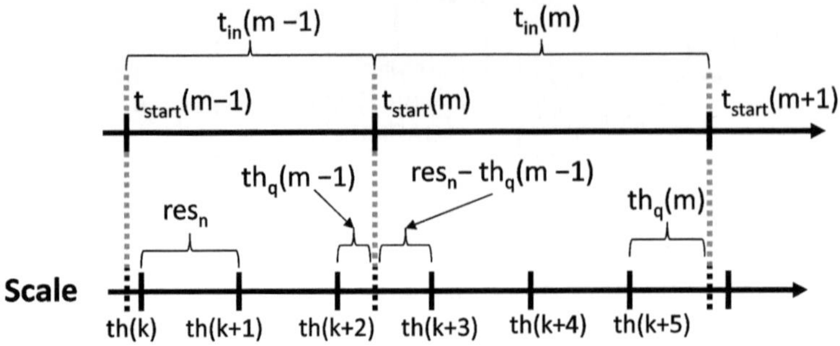

Fig. 3.25 Timing diagram of a GRO-based TDC equivalent to that in Fig. 3.24

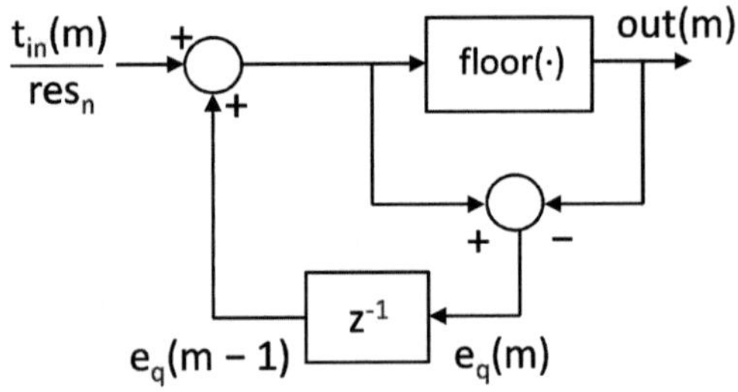

Fig. 3.26 Sigma-delta model of a GRO-based TDC

where floor(·) rounds to the largest integer that is smaller than or equal to the argument. As can be deduced from Fig. 3.25, the duration t_{in} of the time interval that has to be measured is given by:

$$t_{in}(m) = res_n \cdot out(m) + res_n - th_q(m-1) + th_q(m), \tag{3.23}$$

where $out(m)$ is given by Eq. (3.22). By substituting Eq. (3.22) into (3.23), we obtain:

$$e_q(m) = \frac{t_{in}(m)}{res_n} + e_q(m-1) - \text{floor}\left(\frac{t_{in}(m)}{res_n} + e_q(m-1)\right), \tag{3.24}$$

where $e_q(m)$ is given by:

$$e_q(m) = \frac{th_q(m)}{res_n}. \tag{3.25}$$

The block diagram of the first-order sigma-delta modulator in Fig. 3.26 is associated with Eq. (3.24). The quantization error of the sigma-delta modulator in Fig. 3.26 is equal to $e_q(m)$ and is given by Eq. (3.24). The output of the sigma-delta modulator in Fig. 3.26 is equal to the output of an ideal GRO-based TDC at the end of the mth measurement.

Table 3.6 Example of first order noise shaping TDCs with picosecond resolution

Work	Technology	Resolution [ps]
Straayer [32]	0.13 μm	6
Cao [33]	0.13 μm	<1
Lu [34]	90 nm	3.2

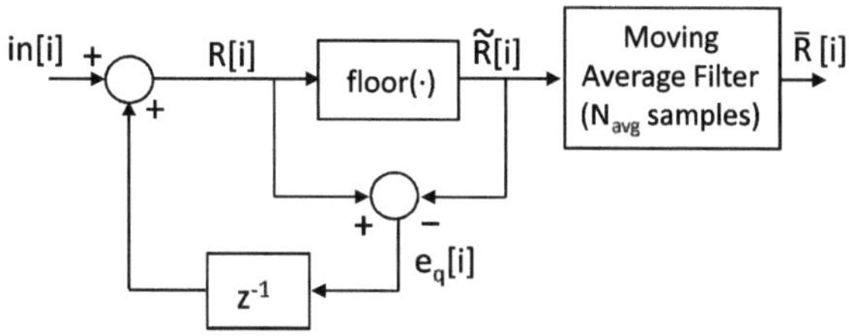

Fig. 3.27 Block diagram of the system "first-order sigma-delta modulator plus moving-average filter"

In an *ideal* GRO-based TDC, we assume that the complete periods of the GRO are constant and equal to the nominal time resolution res_n. The output of an ideal GRO-based TDC is given by Eq. (3.22).

Table 3.6 reports some recent work on GRO-based TDCs with picosecond resolution.

In order to be able to compare the time resolution of a noise shaping TDC with that of other TDC architectures, it is important to point out that the resolution of the system "first-order sigma-delta modulator plus moving-average filter" is different from that of a simple first-order sigma-delta modulator. Hence, we determine the resolution of the system "First-Order Sigma-Delta Modulator plus moving-average filter" shown in Fig. 3.27.

The model in Fig. 3.26 can be seen as a *discrete-time* system that produces measurements at the instants $t_{stop}(m)$. Hence, the sigma-delta modulator in Fig. 3.26 is equivalent to that in Fig. 3.27 when the discrete-time index i is associated with the instants $t_{stop}(m)$. The brackets in Fig. 3.27 emphasize the discrete-time nature of the model.

The output $out_{filt}[i]$ of a moving-average filter is given by:

$$out_{filt}[i] = \frac{1}{N_{avg}} \sum_{g=i-N_{avg}+1}^{i} in_{filt}[g], \tag{3.26}$$

where in_{filt} is the input of the moving-average filter and N_{avg} is the number of samples that are used by the filter. A moving-average filter is often referred to as a "sinc" filter because the Discrete Time Fourier Transform of its impulse response is a sinc function [35].

Before proceeding with the system "First-Order Sigma-Delta Modulator plus Sinc Filter", it is necessary to define the resolution of a first-order sigma-delta modulator. Note that the difference between the output and the input of the sigma-delta modulator in Fig. 3.27 is always included in the interval [−1, 1], as we will demonstrate hereafter. From the block diagram in Fig. 3.27, we can write:

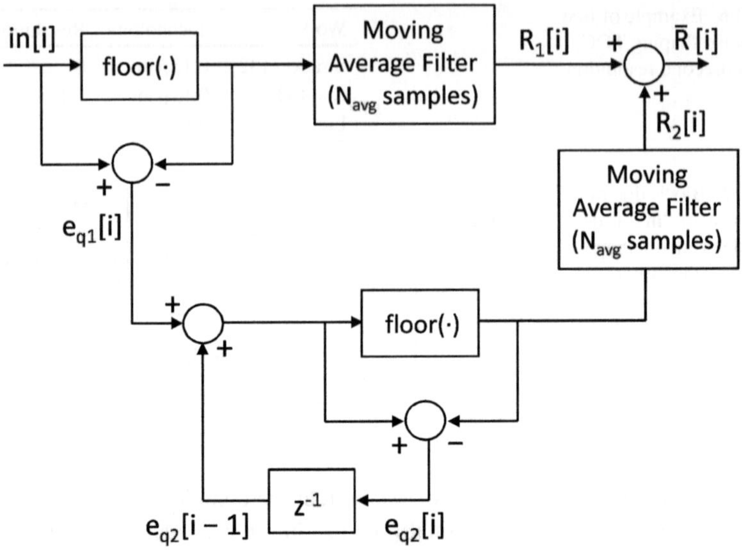

Fig. 3.28 Block diagram of a system equivalent to that in Fig. 3.27

$$in[i] - \tilde{R}[i] = e_q[i] - e_q[i-1]. \tag{3.27}$$

We note that $0 \leq e_q[i] < 1$. Then, $-1 < e_q[i] - e_q[i-1] < 1$ and $-1 < in[i] - R[i] < 1$. We conclude that the output $\tilde{R}[i]$ of the sigma-delta modulator in Fig. 3.27 is a measure of R[i] with resolution equal to two.

In order to determine the resolution of the system "First-Order Sigma-Delta Modulator plus Sinc Filter", consider the block diagram in Fig. 3.28, which is equivalent to that in Fig. 3.27. The output $\overline{R}\,[i]$ of the system in Fig. 3.28 is given by:

$$\overline{R}\,[i] = R_1[i] + R_2[i], \tag{3.28}$$

where

$$R_1[i] = \frac{1}{N_{avg}} \sum_{g=i-N_{avg}+1}^{i} floor\,(in[i]) \tag{3.29}$$

and

$$R_2[i] = \frac{1}{N_{avg}} \sum_{g=i-N_{avg}+1}^{i} \left(e_{q1}[i] - e_{q2}[i] + e_{q2}[i-1]\right). \tag{3.30}$$

Equation (3.30) can be obtained by substituting $in[i] = e_{q1}[g]$ into Eq. (3.27). In Eqs. (3.29) and (3.30), the operator floor $(\cdot)$ rounds to the largest integer less than the argument. Recall that floor $(x) = x - e_q(x)$, where $e_q(x)$ is the quantization error associated with x and $0 \leq e_q(x) < 1$. We can write, from Eq. (3.29):

$$R_1[i] = \frac{1}{N_{avg}} \sum_{g=i-N_{avg}+1}^{i} \big(in[g] - e_{q1}[g]\big). \tag{3.31}$$

Furthermore, we notice that:

$$\frac{1}{N_{avg}} \sum_{g=i-N_{avg}+1}^{i} \big(e_{q2}[g] - e_{q2}[g-1]\big) = \frac{e_{q2}[i] - e_{q2}[i - N_{avg}]}{N_{avg}}. \tag{3.32}$$

Hence, Eq. (3.30) can be simplified to:

$$R_2[i] = \frac{1}{N_{avg}} \sum_{g=i-N_{avg}+1}^{i} e_{q1}[g] - \frac{e_{q2}[i] - e_{q2}[i - N_{avg}]}{N_{avg}}. \tag{3.33}$$

By substituting Eqs. (3.31) and (3.33) into (3.28), we have:

$$\overline{R}[i] = \frac{1}{N_{avg}} \sum_{g=i-N_{avg}+1}^{i} in[g] - \frac{e_{q2}[i] - e_{q2}[i - N_{avg}]}{N_{avg}}. \tag{3.34}$$

From Eq. (3.34), we can express the difference between the averaged input and the output $\overline{R}[i]$ of the sigma-delta modulator as follows:

$$\frac{1}{N_{avg}} \sum_{g=i-N_{avg}+1}^{i} in[g] - \bar{R}[i] = \frac{e_{q2}[i] - e_{q2}[i - N_{avg}]}{N_{avg}}. \tag{3.35}$$

We recall that $0 \leq e_{q2}[i] < 1$, and note that $-1 < e_{q2}[i] - e_{q2}[i - N_{avg}] < 1$. Thus, from Eq. (3.35), we conclude that the average over N_{avg} samples of the output of a first-order sigma-delta modulator is equal to the average over N_{avg} of its input samples with a difference included in the interval $[-res/N_{avg}, res/N_{avg}]$; the latter reduces linearly as N_{avg} increases. Consequently, the resolution of the system "First-Order Sigma-Delta Modulator plus Sinc Filter" in Fig. 3.27 is equal to two divided by N_{avg}, which is N_{avg} times better than that of a first-order sigma-delta modulator without a sinc filter.

3.11 TDC Behavioral Models

Figure 3.29 shows a high-level behavioral model of a TDC based on three blocks that are labelled "Normalized Ideal Duration Measurement", "Quantization" and "Gain", respectively.

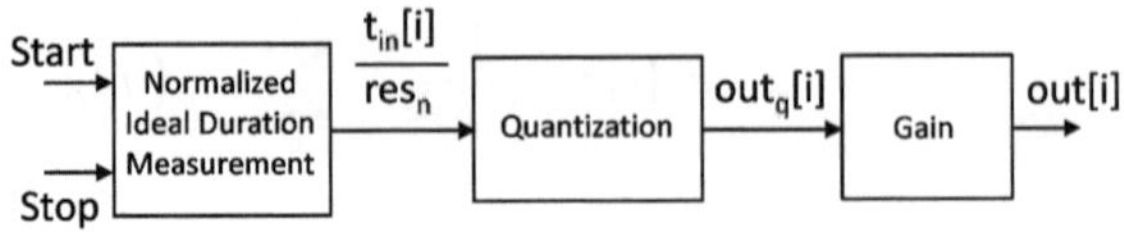

Fig. 3.29 High-level behavioral model of a TDC

The "Normalized Ideal Duration Measurement" block represents the operation of an ideal TDC with infinite time resolution that is able to produce an output equal to the time interval $t_{in}[i]$ that is defined by the edges of the signals "Start" and "Stop". The normalization factor is associated with the time interval that is used as a reference in a TDC and is determined by the TDC timing diagram. The index i denotes the ith measurement and $t_{in}[i]$ represents a discrete time signal. The block "Quantization" in Fig. 3.29 includes the quantization effects that are related to the finite resolution of a TDC. In the previous sections of this chapter, we related the quantization effects in each TDC architecture that has been considered to a timing diagram. The block "Gain" in Fig. 3.29 represents the gain that can be implemented in a TDC by means of a multiplication of digital words.

It is important to note that the index i in Fig. 3.29 is equivalent to the index m that we used to represent the *thresholds* in a first-order sigma-delta TDC. This is not true in the other TDC architectures that we have considered in this chapter. In fact, the difference between two consecutive thresholds $th_{start}(m)$ and $th_{start}(m+1)$ in a first-order sigma-delta TDC is equal to $t_{in}[i]$, as explained in Sect. 3.10. By contrast, the thresholds are independent of $t_{in}[i]$ in the other TDC architectures that we have considered. Thus, the quantization effects in a first-order sigma-delta TDC can be represented by the block diagrams in Fig. 3.30a, b when relating the operation of the TDC to the thresholds $th_{start}(m)$ and the discrete time domain i of the signal $t_{in}[i]$, respectively.

The block "Quantization" in Fig. 3.29 is represented by the floor(·) operator in pulse-shrinking, delay-line-based and oscillator-based TDCs, as shown in Fig. 3.31a and as can be derived by the timing diagrams that we have illustrated in the previous sections. The gain res_n in Fig. 3.31a scales the output of the floor(·) operator to take account into the normalization factor. Finally, K_p represents a gain factor in Fig. 3.31a.

The behavioral model of a GRO-based TDC is shown in Fig. 3.31b. Quantization in a GRO-based TDC is represented by means of the first-order sigma-delta modulator in Fig. 3.31b. The blocks labelled "res_n" and "K_p" Fig. 3.31b are gain factors.

Vernier delay-line-based and Vernier oscillator-based TDCs can be modelled by the block diagram in Fig. 3.32. The index m in Fig. 3.32 is related to the operation of a Vernier TDC between two consecutive measurements. Assume that $t[i]$ and $t[m]$ are the time instants associated with the time indices i and m, respectively. In the interval $[t[i], t[i+1]]$, there are a number of instants $t[m]$ that describe the measurement process of $t_{in}[i]$, as illustrated in Fig. 3.33.

At the instant $t[i]$, the TDC is reset and the value of $e_q[m]$ is set to $t_{in}[i]/res_{Bn}$. The measure is complete at the instant $t[M]$ in the time domain of the index m. At the instant $t[M]$, the output $out_q[M]$ is sampled and held until the instant $t[i+1]$,

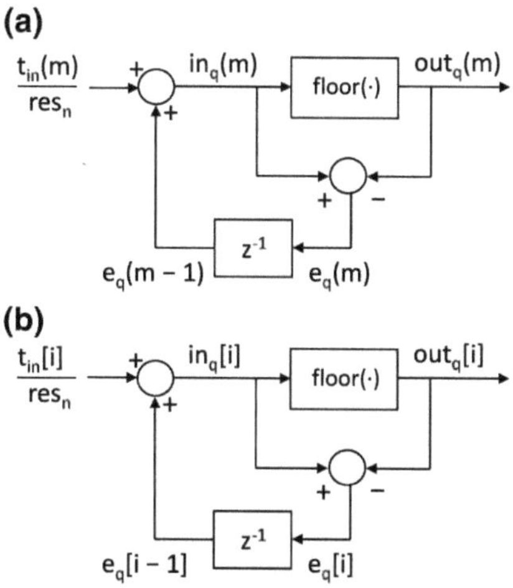

Fig. 3.30 Sigma-delta modulators that are associated with a first-order noise-shaping TDC when relating the operation of the TDC to **a** the thresholds $th_{start}(m)$ and **b** the discrete time domain i of the the signal $t_{in}[i]$

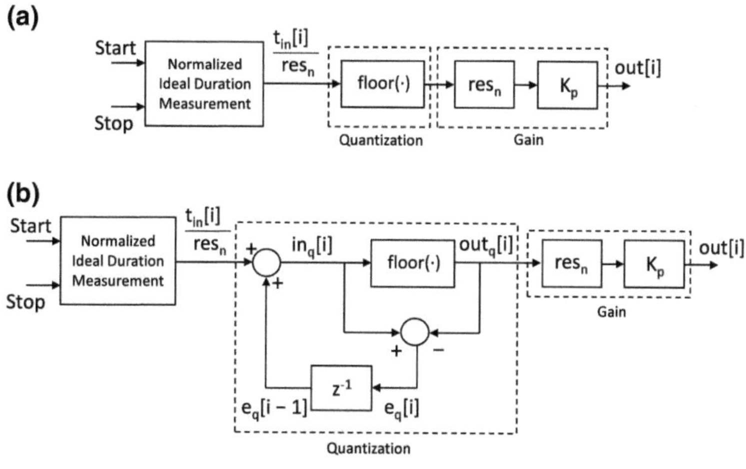

Fig. 3.31 Behavioral models of **a** pulse-shrinking, delay-line-based, oscillator-based and **b** GRO-based TDCs

when the next measurement starts. By sampling and holding, $out_q[m]$ is converted to the time domain of the index i. The block labelled "Sample and Hold" in Fig. 3.32 represents the conversion of out_q from the time domain of the index m to that of the index i.

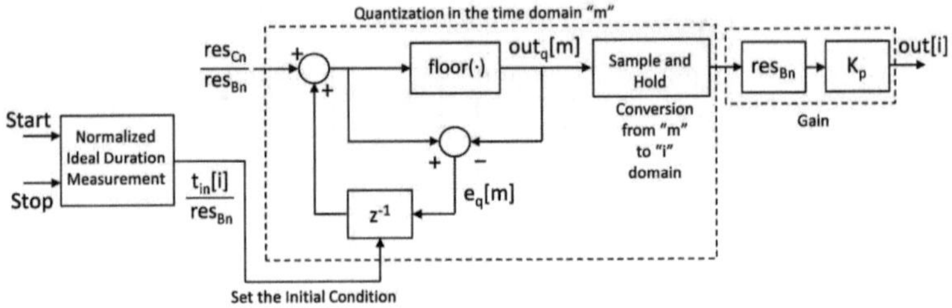

Fig. 3.32 Behavioral model of Vernier delay-line-based and Vernier oscillator-based TDCs

Fig. 3.33 Example timing diagram of the time domains that are defined by the indices *m* and *i* in a Vernier TDC

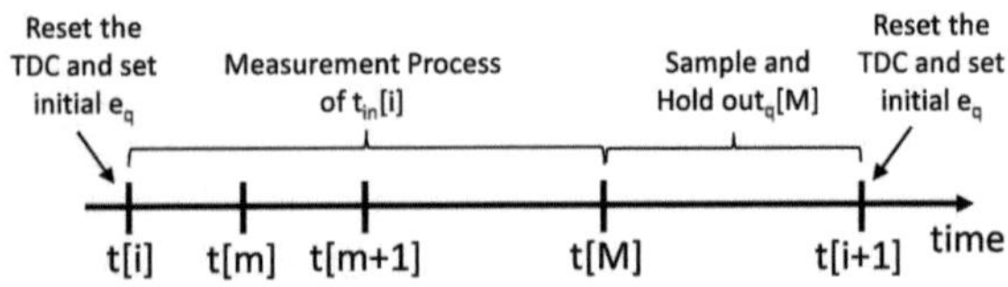

3.12 Comparison

Delay-line-based, oscillator-based and pulse-shrinking TDCs operate with only one time-scale, as we have seen in the previous sections. The operating principle of a single time-scale TDC can be related to that of a flash ADC, as can be seen by comparing the timing diagrams reported in Sects. 3.6–3.9 to that in Sect. 3.2. However, in a flash ADC, all of the comparisons are performed in parallel. In a single-scale TDC, they are performed sequentially. The outputs of a flash ADC and, hence, of a single time-scale TDC can be expressed in terms of the floor$(\cdot)$ operator, as shown in Sect. 3.2.

Vernier delay-line-based, Vernier oscillator-based and GRO-based TDCs use two time scales in order to realize an analog-to-digital conversion. The Vernier method and the operation of a GRO-TDC can be related to a first-order sigma-delta modulator, as shown in Sects. 3.6.2 and 3.10, respectively.

In Vernier delay-line-based and oscillator-based TDCs, consecutive measurements are independent. The duration of the time interval that has to be measured is related to the *initial* value of the quantization error of a first-order sigma-delta modulator, as explained in Sect. 3.6.2. By contrast, consecutive measurements in a GRO-based TDC are *dependent*; the duration of the time interval that has to be measured is related to the input of a first-order sigma-delta modulator.

Figure 3.34 shows a summary of the TDC architectures that we have considered in this chapter and their time resolutions.

The GRO-based TDC is the *only* TDC architecture with *dependent consecutive* measurements, as can be seen from Fig. 3.34. This feature has an important impact on the time resolution of the GRO-based TDC. In fact, by processing consecutive measurements of a GRO-based TDC, it is possible to achieve higher time resolution. In order to illustrate the relationship between the resolution and first-order noise-

Single Time-Scale TDCs (equivalent to flash ADCs)				
Architecture	Delay-line based TDC	Oscillator-based TDC	Mixed oscillator- and delay-line-based TDC	Pulse-shrinking TDC
Nominal Time Resolution	Duration of a delay	Duration of a period	Duration of a delay	Difference between the durations of consecutive pulses

Dual Time-Scale TDCs (equivalent to sigma-delta ADCs)			
Consecutive Measures	**Independent**		**Dependent**
Architecture	Vernier delay-line-based TDC	Vernier oscillator- based TDC	GRO-based TDC
Nominal Time Resolution	Difference between the durations of two delays	Difference between the durations of two periods	Duration of a period*

*The time resolution of the system "GRO-based TDC plus moving average filter" is equal to two times the duration of a period divided by the number of samples that are used by the moving average filter.

Fig. 3.34 TDC architectures that we considered in this chapter and their time resolutions

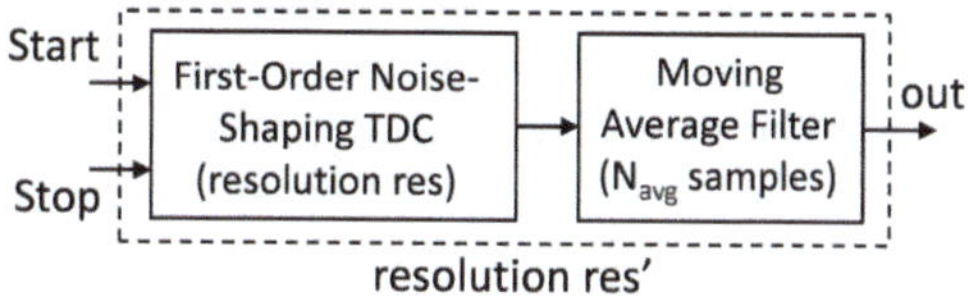

Fig. 3.35 System "first-order noise-shaping TDC plus moving-average filter"

shaping, let us consider the system "first-order noise-shaping TDC plus moving-average filter" that is shown in Fig. 3.35.

The time resolution of the TDC and the number of samples of the moving-average filter in Fig. 3.35 are res and N_{avg}, respectively. As demonstrated in Sect. 3.10, the resolution of the system "first-order noise-shaping TDC plus moving-average filter" is $res' = res/N_{avg}$. The time resolution of an *ideal* "first-order noise-shaping TDC plus moving-average filter" can be increased *without limit* by *digitally* processing more samples of its output signal. However, the time that the step response of a moving-average filter takes to reach steady state increases with N_{avg}. Hence, there is a tradeoff between the resolution and the time required to reach steady state in a "first-order noise-shaping TDC plus moving-average filter". Moreover, tones might be present in the output of a first-order noise-shaping TDC [36]. The frequencies and amplitudes of the tones in an ideal first-order sigma-delta modulator are predictable [36] and can be taken into account in the design of the system "first-order noise-shaping TDC plus moving-average filter".

Note that the output of a TDC is *always* equal to zero when the input is smaller than the time resolution in all of the other TDC architectures that we have considered. Hence, the system "zeroth order noise-shaping TDC plus moving-average filter" produces a sequence of zeros, regardless of the value of N_{avg}, when the duration of the time interval that has to be measured is smaller than the time resolution of the TDC. It is important to note that, when dither is added to the input of a quantizer, the effective resolution of the system "quantizer plus moving average filter" increases with N_{avg}, as will be demonstrated in Chap. 5. However, we will also show in Chap. 5 that the precision of the system "sigma-delta modulator plus moving average filter" is higher than that of a "quantizer plus moving average filter" when the same dither is added to the input.

We conclude that a GRO-based TDC, or a general "noise-shaping TDC", is an interesting architecture for digital systems that allows a tradeoff between resolution in time and operating speed.

3.13 Conclusion

In this chapter, we have summarized the operating principles of the main TDC architectures that comprise delay-line-based, oscillator-based, pulse-shrinking, Vernier-delay-line-based, Vernier-oscillator-based and GRO-based TDCs. We showed that all architectures can be represented as equivalent single-scale or dual-scale analog-to-digital converters. Moreover, we demonstrated analytically that a Vernier TDC is a particular case of a sigma-delta TDC. We showed that a GRO-based TDC, or a general "noise-shaping TDC", is an interesting architecture for digital systems that allows a tradeoff between time resolution and operating speed.

References

1. J. Kalisz, "Review of Methods for Time Interval Measurements withPicosecond Resolution," *Metrologia*, vol. 41, no. 1, pp.1732, Feb. 2004.
2. Maxim, "Understanding Flash ADCs," *Application Note 810*,Sep. 2010.
3. R. B. Staszewski, S. Vemulapalli, P. Vallur, J. Wallberg, P. T. Balsara, "1.3 V 20 ps Time-to-Digital Converter for Frequency Synthesis in 90-nm CMOS," *IEEE Trans. Circuits and Sys. II: Express Briefs*, vol. 53, no. 3, Mar. 2006.
4. M. Terabe, A. Sekiya, T. Yamada, and A. Fujimaki, "Timing Jitter Measurement in Single-Flux-Quantum Circuits Based on Time-to-Digital Converters With High Time-Resolution," *IEEE Trans. Applied Superconductivity*, vol. 17, no. 2, Jun. 2007.
5. A. S. Yousif, J. W. Haslett, "A Fine Resolution TDC Architecture for Next Generation PET Imaging," *IEEE Trans. Nuclear Science*, vol. 54, no. 5, Oct. 2007.
6. M. Lee, A. A. Abidi, "A 9 b, 1.25 ps Resolution Coarse-Fine Time-to-Digital Converter in 90 nm CMOS that Amplifies a Time Residue" *IEEE J. of Solid-State Circuits*, vol. 43, no. 4, Apr. 2008.
7. M. Lee, M. E. Heidari, A. A. Abidi, "A Low-Noise Wideband Digital Phase-Locked Loop Based on a CoarseFine Time-to-Digital Converter With Subpicosecond Resolution," *IEEE J. Solid-State Circuits*, vol. 44, no. 10, Oct. 2009.
8. J. Wang, S. Liu, Q. Shen, H. Li, Q. An, "A Fully Fledged TDC Implemented in Field-Programmable Gate Arrays," *IEEE Trans. Nuclear Science*, vol. 57, no. 2, Apr. 2010.

9. P. Chen, C.-C. Chen, Y.-S. Shen, "A Low-Cost Low-Power CMOS Time-to-Digital Converter Based on Pulse Stretching," *IEEE Trans. Nuclear Science*, vol. 53, no. 4, Aug. 2006.
10. J. Song, Q. An, S. Liu, "A High-Resolution Time-to-Digital Converter Implemented in Field-Programmable-Gate-Arrays," *IEEE Trans. Nuclear Science*, vol. 53, no. 1, Feb. 2006.
11. J.-P. Jansson, A. Mantyniemi, J. Kostamovaara, "A CMOS Time-to-Digital Converter With Better Than 10 ps Single-Shot Precision,"*IEEE J. Solid-State Circ.*, vol. 41, no. 6, Jun. 2006.
12. J. Nissinen, I. Nissinen, J. Kostamovaara, "Integrated Receiver Including Both Receiver Channel and TDC for a Pulsed Time-of-Flight Laser Rangefinder With cm-Level Accuracy," *IEEE J. Solid-State Circuits*, vol. 44, no. 5, May 2009.
13. J.-P. Jansson, A. Mntyniemi, J. Kostamovaara, "Synchronization in a Multilevel CMOS Time-to-Digital Converter," *IEEE Trans. Circuits Sys. I: Regular Papers*, vol. 56, no. 8, Aug. 2009.
14. Encyclopaedia Britannica Online "Vernier Caliper". Available athttp://www.britannica.com/EBchecked/topic/626328/vernier-caliper
15. K. Nose, M. Kajita, M. Mizuno, "A 1-ps Resolution Jitter-Measurement Macro Using Interpolated Jitter Oversampling," *IEEE J. Solid-State Circuits*, vol. 41, no. 12, Dec. 2006.
16. S. Henzler, S. Koeppe, D. Lorenz, W. Kamp, R. Kuenemund, D. Schmitt-Landsiedel, "A Local Passive Time Interpolation Concept for Variation-Tolerant High-Resolution Time-to-Digital Conversion," *IEEE J. Solid-State Circuits*, vol. 43, no. 7, Jul. 2008.
17. E. Temporiti, C. Weltin-Wu, D. Baldi, R. Tonietto, F. Svelto, "A 3 GHz Fractional All-Digital PLL With a 1.8 MHz Bandwidth Implementing Spur Reduction Techniques," *IEEE J. Solid-State Circuits*, vol. 44, no. 3, Mar. 2009.
18. A. M. Amiri, M. Boukadoum, A. Khouas, "A Multihit Time-to-Digital Converter Architecture on FPGA," *IEEE Trans. Instrum. Meas.*, vol. 58, no. 3, Mar. 2009.
19. N. Xing, W.-Y. Shin, D.-K. Jeong and S. Kim, "High-resolution time-to-digital converter utilising fractional difference conversion scheme," *Electronics Letters*, vol. 46 no. 6, Mar. 2010.
20. R. Rashidzadeh, M. Ahmadi, W. C. Miller, "An All-Digital Self-Calibration Method for a Vernier-Based Time-to-Digital Converter," *IEEE Trans. Instrum. Meas.*, vol. 59, no. 2, Feb. 2010.
21. L. Vercesi, A. Liscidini, R. Castello, "Two-Dimensions VernierTime-to-Digital Converter," *IEEE J. Solid-State Circuits*, vol. 45, no. 8, Aug. 2010.
22. P. Chen, C.-C. Chen, J.-C. Zheng, Y.-S. Shen, "A PVT Insensitive Vernier-Based Time-to-Digital Converter With Extended Input Range and High Accuracy," *IEEE Trans. Nuclear Science*, vol. 54, no. 2, Apr. 2007.
23. A. K. M. K. Mollah, R. Rosales, S. Tabatabaei, J. Cicalo, A. Ivanov, "Design of a Tunable Differential Ring Oscillator With Short Start-Up and Switching Transients," *IEEE Trans. Circuits Sys. I: Regular Papers*, vol. 54, no. 12, Dec. 2007.
24. S. S. Junnarkar, P. OConnor, P. Vaska, R. Fontaine, "FPGA-Based Self-Calibrating Time-to-Digital Converter for Time-of-Flight Experiments," *IEEE Trans. Nuclear Science*, vol. 56, no. 4, Aug. 2009.
25. J. Yu, F. F. Dai, R. C. Jaeger, "A 12-Bit Vernier Ring Time-to-Digital Converter in 0.13 μ m CMOS Technology," *IEEE J. Solid-State Circuits*, vol. 45, no. 4, Apr. 2010.
26. P. Chen, P.-Y. Chen, J.-S. Lai, Y.-J. Chen, "FPGA Vernier Digital-to-Time ConverterWith 1.58 ps Resolution and 59.3 Minutes Operation Range," *IEEE Trans. Circuits Sys. I: Regular Paper*, vol. 57, no. 6, Jun. 2010.
27. C.-C. Chen, P. Chen, C.-S. Hwang, W. Chang, "A Precise Cyclic CMOSTime-to-Digital Converter With Low Thermal Sensitivity," *IEEETrans. Nuclear Science*, vol. 52, no. 4, Aug. 2005.
28. P. Chen, S.-I. Liu, and Jingshown Wu, "A CMOS Pulse-Shrinking DelayElement For Time Interval Measurement" *IEEE Trans. CircuitsSys. II: Analog and Digital Signal Processing*, VOL. 47, NO. 9, Sep.2000.
29. K. Karadamoglou, N. P. Paschalidis, E. Sarris, N. Stamatopoulos, G.Kottaras, V. Paschalidis, "An 11-bit High-Resolution andAdjustable-Range CMOS Time-to-Digital Converter for Space ScienceInstruments" *IEEE J. Solid-State Circuits*, vol. 39, no. 1,Jan. 2004.
30. Y. Liu, U. Vollenbruch, Y. Chen, C. Wicpalek, L. Maurer, Z. Boos,R. Weigel, "Multi-stage Pulse Shrinking Time-to-Digital Converterfor Time Interval Measurements," *Proceedings of the 10thEuropean Conference on Wireless Technology (ECWT 2007)* pp. 347–350.

31. R. Szplet, K. Klepacki, "An FPGA-Integrated Time-to-DigitalConverter Based on Two-Stage Pulse Shrinking," *IEEE Trans. Instrum. Meas.*, vol. 59, no. 6, Jun. 2010.
32. M. Z. Straayer, M. H. Perrott, "A Multi-Path Gated Ring OscillatorTDC With First-Order Noise Shaping," *IEEE J. Solid-StateCircuits*, vol. 44, no. 4, Apr. 2009.
33. Y. Cao, P. Leroux, W. De Cock, M. Steyaert "A 1.7mW 11b 1-1-1 MASH $\Delta\Sigma$ Time-to-Digital Converter" *Solid-State Circuits Conference Digest of Technical Papers (ISSCC 2011)*, pp. 480–482, Feb. 2011.
34. P. Lu, A. Liscidini, P. Andreani "A 3.6 mW, 90 nm CMOSGated-Vernier Time-to-Digital Converter With an EquivalentResolution of 3.2 ps" *IEEE J. Solid-State Circuits*, vol. 99,May. 2012.
35. S. Haykin, B. Van Veen, "Signals and Systems," Second Edition,*John Wiley & Sons, Inc.,* 2003.
36. R. Schreier, G. C. Temes, "Understanding Delta-Sigma Data Converters", *IEEE Press, John Wiley & Sons, Inc.*, 2005.

Chapter 4
Analytical Predictions of Phase Noise in ADPLLs

4.1 Introduction

In this chapter, we will derive analytical predictions of the phase noise in TDC-based and accumulator-based ADPLLs with lth-order noise shaping TDCs and DCO driven by a sigma-delta modulator. The architectures of TDC-based and accumulator-based ADPLLs are shown in Fig. 4.1.

In order to derive analytical predictions of the ADPLL's phase noise, we first explain how to calculate the phase noise of the DCO when its input is known. Then we will derive linear models associated with the building blocks in an ADPLL. We will use the linear models of the building blocks of an ADPLL to develop a linear model of the full ADPLL. Analytical predictions of the ADPLL's phase noise will be based on the linear ADPLL models.

Finally, we compare Matlab simulations and analytical predictions for an example TDC-based ADPLL architecture.

4.2 Phase Noise

The power spectral density of an ideal *noiseless* oscillator with a free-running frequency f_0 is an impulse at the frequency f_0, as shown in Fig. 4.2a. The power spectral density $L(f)$ of an *actual noisy* oscillator is a Lorentzian spectrum centred at the frequency f_0 [1], as illustrated in Fig. 4.2b.

$L(f)$ can be calculated from the power spectral density of the *phase deviation* of the output of an oscillator. The phase deviation $\Delta\Phi(t)$ is defined as the phase difference between the phase of the noiseless oscillator and the actual noisy phase, as illustrated in Fig. 4.3.

$L(f)$ is given by [1]:

$$L(f_m) = \frac{S_{\Delta\Phi}(f_m)}{2} \tag{4.1}$$

F. Brandonisio and M. P. Kennedy, *Noise-Shaping All-Digital Phase-Locked Loops*, Analog Circuits and Signal Processing, DOI: 10.1007/978-3-319-03659-5_4,

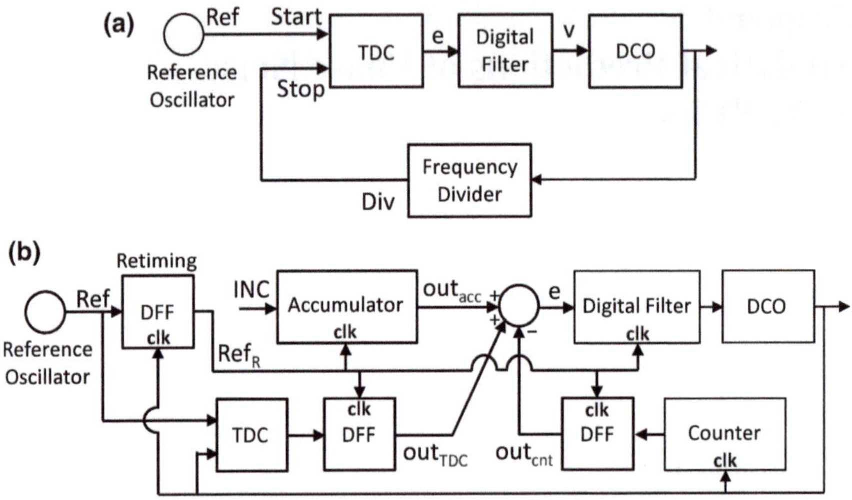

Fig. 4.1 Architectures of **a** TDC-based and **b** accumulator-based ADPLLs

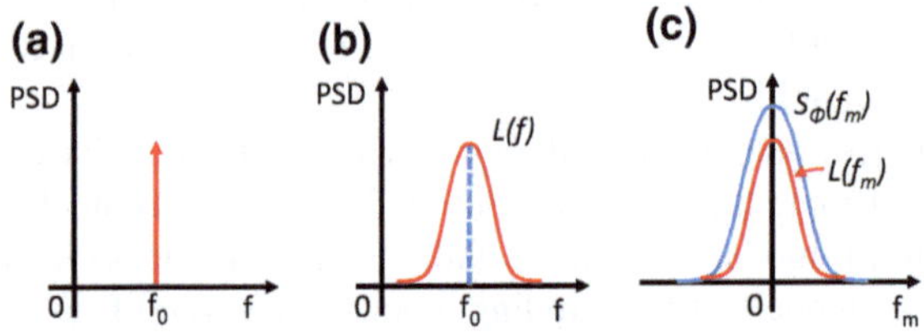

Fig. 4.2 Example power spectral densities of the outputs signals of **a** an ideal noiseless oscillator and **b** an actual oscillator. **c** Example phase noise $L(f_m)$ and power spectral density $S_{\Delta\Phi}(f_m)$ of the phase deviation of an actual oscillator

Fig. 4.3 Example phase deviation $\Delta\Phi(t)$, ideal and perturbed phases of an oscillator

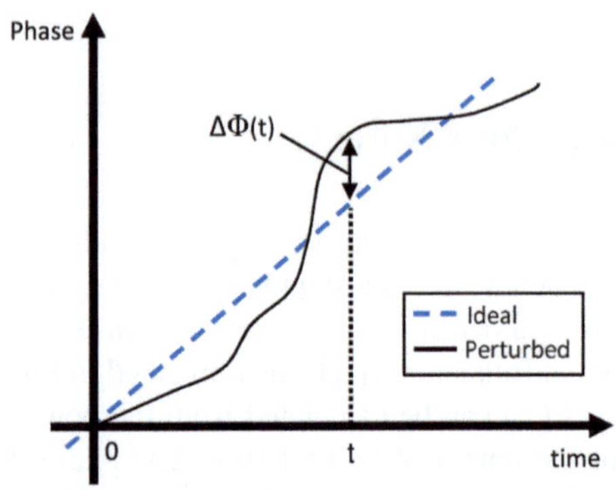

where $S_{\Delta\Phi}(f_m)$ is the power spectral density of the phase deviation of the output signal of the oscillator and $f_m = f - f_0$. Equation (4.1) will be analysed in detail in Chap. 7.

DCOs and VCOs are oscillators in which the frequency, and thus the phase, is related to the input. Thus, it is possible to relate the phase deviation of DCOs and VCOs to their inputs. A DCO can be seen as a VCO with quantization effects. Hence, we will apply to a DCO the existing techniques that are used to determine the phase noise of a VCO from its input. The simulator CppSim calculates the phase noise of a VCO from samples of the VCO input by means of the Matlab code in Listing 4.1. A similar procedure is shown in [2]. The procedures used by CppSim and shown in [2] are in agreement with those that will be described in detail in Chap. 7.

```
num_segments = 20;

window_length = floor(length(t)/num_segments);

Kv = 1; % VCO gain (Hz/V) - this is one for noiseout
    from CppSim
phase = filter(Ts*2*pi*Kv,[1 -1],vin-mean(vin));
%  calculate L(f)
[Pxx,f] =
    pwelch(phase,window_length,[],[],1/Ts,'twosided');

Pxx_db = 10*log10(Pxx); semilogx(f,Pxx_db);
```

Listing 4.1 *Matlab script for calculating the phase noise (in dBc/Hz) of a VCO with a gain* `Kv` *from the VCO input voltage* `vin` *sampled at a frequency 1/Ts.*

Listing 4.1 calculates the DCO frequency deviation from the VCO input voltage `vin` sampled at the frequency `1/Ts`[1] by subtracting the mean value from `vin` and multiplying the result by the gain Kv of the VCO. The phase deviation is obtained by integrating the frequency deviation with the function "`filter`(·)". It is important to notice that the integration is necessary to take into account the accumulation of the phase deviation that occurs in an oscillator. This process is referred to as accumulating jitter in the time domain [2]. The function "`pwelch`(·)" is used to obtain the power spectral density of the phase noise in dBc/Hz. We will use Listing 4.1 to calculate the phase noise of a DCO in an ADPLL.

In order to compare analytical predictions and simulations, it is necessary to take into account the operations of Listing 4.1 in the analytical predictions. The analytical prediction of the phase noise $L(f_m)$ of a DCO in an ADPLL can be obtained from a linear model of the ADPLL. In fact, from the linear model of an ADPLL it is possible to write an analytical expression of the power spectral density of the DCO input when all the noise sources in the ADPLL are known. The analytical predictions of the ADPLL phase noise can be obtained by applying to the power spectral density of the DCO input the operations in the frequency domain that are equivalent to those in Listing 4.1.

---

[1] An array `vin` including the samples and the variable `Ts` have to be present in the Matlab workspace in order to use Listing 4.1.

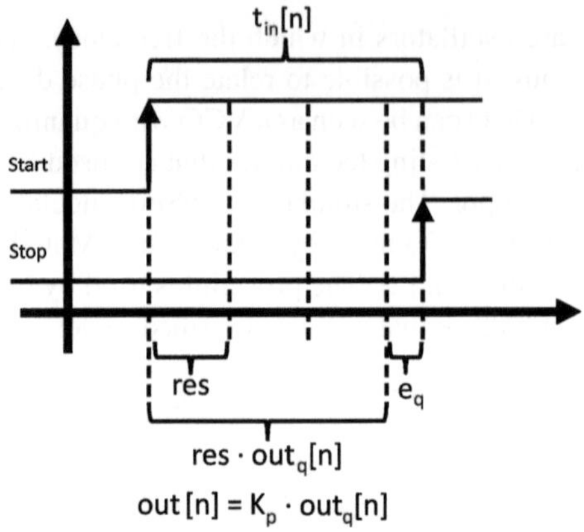

Fig. 4.4 Concept of a time-to-digital converter

In the next sections, we will show how to associate linear models and transfer functions to the building blocks of an ADPLL. We will use the models of the building blocks to determine analytically (i) the power spectral density of the noise at the DCO input and, (ii) the ADPLL's phase noise.

4.3 Models of Time-to-Digital Converters

As mentioned in Chap. 3, a TDC measures the duration of the time interval between two events that can be associated with the rising-edges of two signals *Start* and *Stop*.

The output $out[n]$ of a TDC with finite resolution res_{TDC} is proportional to the number $out_q[n]$ of integer reference time intervals with duration res_{TDC} that are included in the time interval with duration $t_{in}[n]$ that has to be measured, as illustrated in Fig. 4.4.

Figure 4.5a shows the block diagram of a TDC with finite resolution res_{TDC}. The block labelled "Ideal Duration Measurement" in Fig. 4.5a produces a measure of the duration of $t_{in}[n]$ with infinite resolution. The block labelled "K_p" in Fig. 4.5a is a gain equal to K_p. The output of the block labelled "Quantization" in Fig. 4.5a is given by:

$$out_q[n] = \text{floor}\left(\frac{t_{in}[n]}{res_{TDC}}\right), \tag{4.2}$$

where res_{TDC} is the resolution of the TDC. Equation (4.2) can be written as follows:

$$out_q[n] = \frac{t_{in}[n]}{res_{TDC}} - e_{qTDC}[n], \tag{4.3}$$

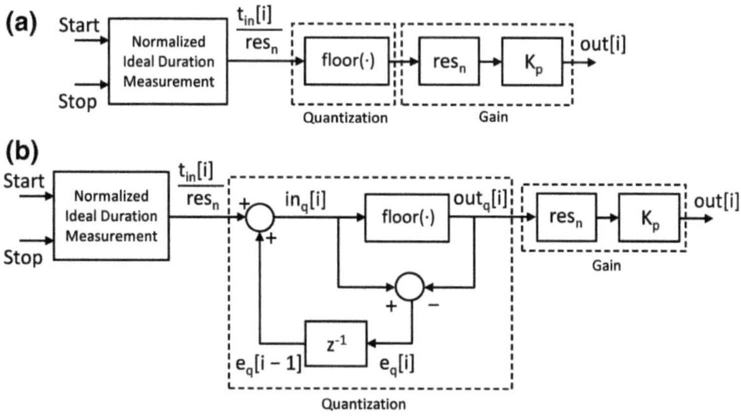

Fig. 4.5 Block diagram of TDCs **a** without noise-shaping and **b** with first-order noise-shaping of the quantization error

where $e_{qTDC}[n]$ is the quantization error and floor$(\cdot)$ rounds to the largest integer less than the argument. The quantization error e_{qTDC} is given by:

$$e_{qTDC}[n] = \frac{t_{in}[n]}{res_{TDC}} - \text{floor}\left(\frac{t_{in}[n]}{res_{TDC}}\right). \tag{4.4}$$

A TDC with finite resolution res_{TDC} and first-order noise-shaping of the quantization error is associated with the block diagram in Fig. 4.5b, as demonstrated in Chap. 3. The model in Fig. 4.5b includes a first-order sigma-delta modulator. The output $out_q[n]$ of a first-order sigma-delta modulator with input t_{in}/res_{TDC} is given by:

$$out_q[n] = \text{floor}\left(\frac{t_{in}}{res_{TDC}} + e_{qTDC}[n-1]\right). \tag{4.5}$$

$e_{qTDC}[n]$ in Eq. 4.5 is given by:

$$e_{qTDC}[n] = \frac{t_{in}}{res_{TDC}} + e_{qTDC}[n-1] - \text{floor}\left(\frac{t_{in}}{res_{TDC}} + e_{qTDC}[n-1]\right). \tag{4.6}$$

Equation (4.4) defines the *additive noise model* that is represented in Fig. 4.6a. The additive noise model applied to a first-order sigma-delta is shown in Fig. 4.6b. The model in Fig. 4.6b can be simplified to that in Fig. 4.7a.

The model in Fig. 4.7a can also be obtained from Eq. (4.5) by means of the z-transform. In the z-domain, Eq. (4.5) becomes:

$$out_q(z) = \frac{t_{in}(z)}{res_{TDC}} + H_{nsTDC}(z)e_{qTDC}(z), \tag{4.7}$$

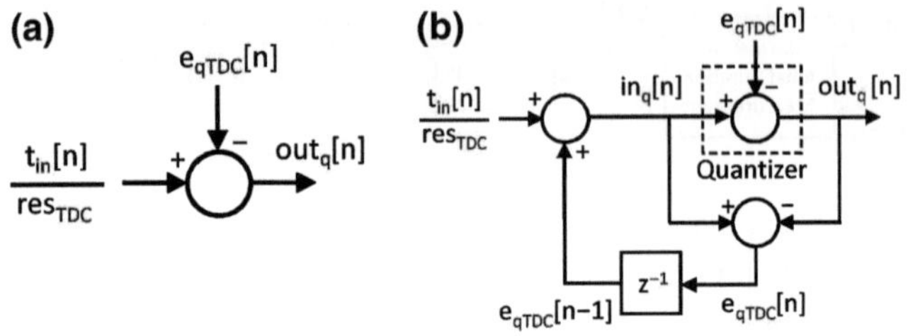

Fig. 4.6 Additive noise model applied to **a** a quantizer and **b** a first-order sigma-delta modulator

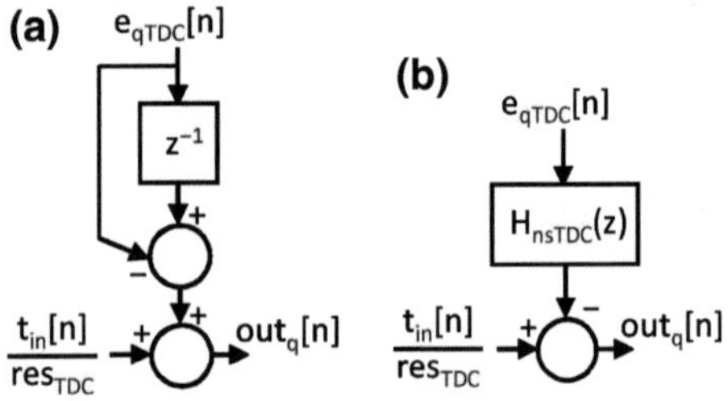

Fig. 4.7 Discrete time models of **a** a first-order noise-shaping and **b** an lth-order noise-shaping TDC

where $H_{nsTDC}(z)$ is the transfer function of the quantization error and is given by:

$$H_{nsTDC}(z) = (1 - z^{-1}). \tag{4.8}$$

An equation with the same form as Eq. (4.7) can be written for a general lth-order noise shaping TDC. In fact, the z-domain output $out_q(z)$ of an lth-order sigma-delta modulator with input $in_q(z)$ is given by [3]:

$$out_q(z) = in_q(z) + (1 - z^{-1})^l e_{qTDC}(z). \tag{4.9}$$

By substituting $in(z) = t_{in}(z)/res_{TDC}$ into Eq. (4.9) we have that the output $out_q(z)$ of an lth-order noise shaping TDC is given by:

$$out_q(z) = \frac{t_{in}(z)}{res_{TDC}} + (1 - z^{-1})^l e_q(z). \tag{4.10}$$

where $H_{nsTDC}(z)$ is the transfer function of the quantization error and is given by:

$$H_{nsTDC}(z) = (1 - z^{-1})^l. \tag{4.11}$$

From Eq. (4.10), we obtain the general representation of an lth-order noise shaping TDC that is shown in Fig. 4.7b.

Equation (4.10) with $l = 0$ gives the output of a simple quantizer in the z-domain. We conclude that a TDC without noise shaping, such as that shown in Fig. 4.5a, can be understood as a *zeroth-order* noise-shaping TDC.

Equation (4.10) can be used to determine analytically the PSD of the output $out_q[n]$ of an lth-order noise shaping TDC.

Since the model in Fig. 4.7b is linear, and assuming that the t_{in}/res_{TDC} and $e_q[n]$ are independent,[2] the power spectral density S_{out} of out_q is given by:

$$S_{out}(z) = S_{in}(z) + |H_{nsTDC}|^2 \, S_{eq}(z) \tag{4.12}$$

where $S_{in}(z)$ and $S_{eq}(z)$ are the power spectral densities of $t_{in}[n]/res_{TDC}$ and e_q, respectively, and $|\cdot|$ is the modulus operator.

Finally, it is convenient to express H_{nsTDC} as a function of the angular frequency Ω in order to compare analytical predictions with simulations. We recall that the z and s variables can be related by means of [5]:

$$z = e^{sT_s}, \tag{4.13}$$

where T_s is the sampling time associated with the z-domain. By substituting Eq. (4.13) into Eq. (4.30) we have:

$$H_{nsTDC}(e^{sT_s}) = (1 - e^{-sT_s})^l. \tag{4.14}$$

By substituting $s = j\Omega$ into Eq. (4.14) and using Euler's formula, we have:

$$\left| H_{nsTDC}(e^{sT_s}) \right|^2 = (2 - 2\cos(T_s\Omega))^l, \tag{4.15}$$

4.4 DCO Model

At the behavioral level, a DCO can be seen as a case of a VCO with quantized input [6]. In fact, a VCO can be viewed as a DCO with infinite resolution. Hence, we can start from the techniques to realize VCO behavioral models [2, 7] in order to obtain the model of a DCO. The simplified behavioral model that relates the frequency f_{VCO} of a noiseless ideal VCO is given by [2, 7]:

$$f_{VCO} = f_0 + K_v in, \tag{4.16}$$

where in is the input voltage of the VCO. A noisy VCO can be represented as an ideal noiseless VCO with equivalent additive noise at the VCO input [2, 7]. Hence, the model represented by Eq. (4.16) for a noisy VCO becomes:

$$f_{VCO} = f_0 + K_v(in + in_{noise}), \tag{4.17}$$

[2] The input and the quantization error in a quantizer are dependent. However, it is common [3, 4] to assume that they are independent in order to simplify the analysis.

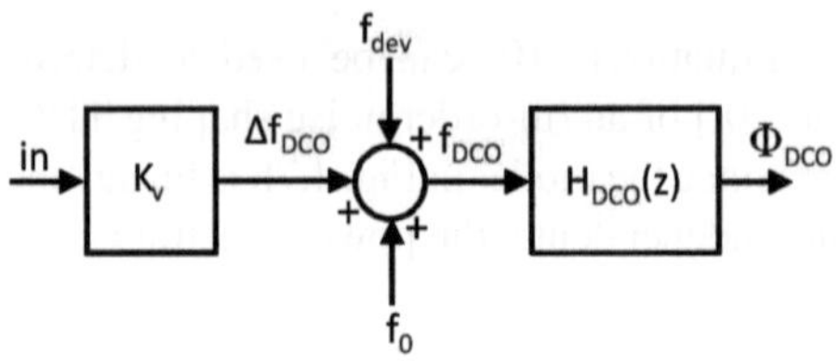

Fig. 4.8 Behavioral model of a DCO with infinite resolution and additive noise expressed as frequency deviations f_{dev}

where in_{noise} is additive input noise. From Eq. (4.18), we can write:

$$f_{VCO} = f_0 + K_v in + f_{dev}, \tag{4.18}$$

where the noise is represented with an additive frequency deviation f_{dev} given by $f_{dev} = in + in_{noise}$. The VCO model given by Eq. (4.18) can be seen as the model of a noisy DCO with *infinite* resolution. Hence, the behavioural models of a noiseless and noisy ideal DCO with infinite resolution can be obtained by substituting the DCO frequency f_{DCO} for the VCO frequency f_{VCO} in Eqs. (4.16) and (4.18), respectively.

Figure 4.8 shows a block diagram associated with Eq. (4.18) in which f_{DCO} replaces f_{VCO}. The signal Φ_{DCO} in Fig. 4.8 is the phase of the VCO, which is the integral of the frequency f_{VCO}. The block labelled $H_{DCO}(z)$ in Fig. 4.8 represents a discrete-time transfer function $H_{DCO}(z)$ associated with an ideal continuous-time integrator. We assume that the discrete time domain associated with $H_{DCO}(z)$ is determined by the instants at which the rising edges of the VCO occur. In fact, we also assume that the input in and the noise f_{dev} in Fig. 4.8 are discrete-time representations of their continuous-time equivalent signals. In the next sections we will determine discrete-time ADPLL models associated with the instants at which the rising edges of the *reference oscillator* occur. The time domain of the edges of the reference oscillator is different from that of the DCO edges. Thus, it is important to determine the DCO model in the time domain of the edges of the reference oscillator. We notice that the s-domain transfer function $H_{DCO}(s)$ of an ideal integrator is given by:

$$H_{DCO}(s) = \frac{1}{s}, \tag{4.19}$$

A discrete-time representation of an $H_{DCO}(s)$ can be obtained by applying the bilinear transform [8] to Eq. (4.19). We recall that the bilinear transform is based on the following relationship [8]:

$$s = \frac{2\,(z-1)}{T\,(z+1)}, \tag{4.20}$$

where T is the duration of the time interval between consecutive discrete-time events. By setting T equal to the period of the reference oscillator in Eq. (4.20) and by substituting the result into Eq. (4.19), we obtain that $H_{DCO}(z)$ in the time domain of the edges of the reference oscillator is given by:

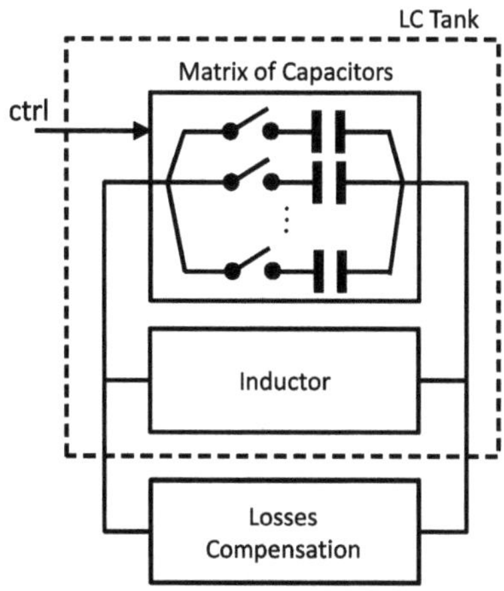

Fig. 4.9 Block diagram of a DCO based on an LC-tank and an active circuit that compensates the energy lost in the LC-tank

$$H_{DCO}(s) = \frac{T(z+1)}{2(z-1)}, \tag{4.21}$$

In order to obtain the model of a noisy DCO with *finite* resolution from Fig. 4.8, we assume a DCO based on an LC tank [9] in which the capacitance is realized by means of a matrix of capacitors, as shown in Fig. 4.9.

The blocks labelled "Inductor" and "Matrix of Capacitors" in Fig. 4.9 represent the LC-tank. The block labelled "Loss Compensation" in Fig. 4.9 represents an active circuit that compensates for the energy that is lost in the LC-tank which has non-ideal (lossy) inductor and capacitors. The input $ctrl$ in Fig. 4.9 is a digital word that determines the states of the switches included in the block "Matrix of Capacitors". The switches connect the capacitors to the terminals of the block "Matrix of Capacitors". The capacitance of the LC-tank is determined by the states of the switches in the matrix of capacitors. We assume that the capacitance of a single capacitor is equal to the capacitive resolution res_{DCO} of the DCO. The free-running oscillation frequency of an LC DCO is given approximately by:

$$f_0 = \frac{1}{\sqrt{LC}}, \tag{4.22}$$

where L and C are the inductance of the inductor and capacitance of the matrix of capacitors in the LC tank when the input of the DCO is equal to zero. When the DCO input is different from zero, the capacitance of the matrix of capacitors is equal to $C + \Delta C$. In order to find the analytical relationship between ΔC and the DCO input, we first consider a DCO with infinite resolution. Then, we include the quantization effects in the DCO model. When the capacitance in the LC-tank is $C + \Delta C$, the oscillation frequency f_{DCO} of a noiseless DCO with infinite resolution is given by:

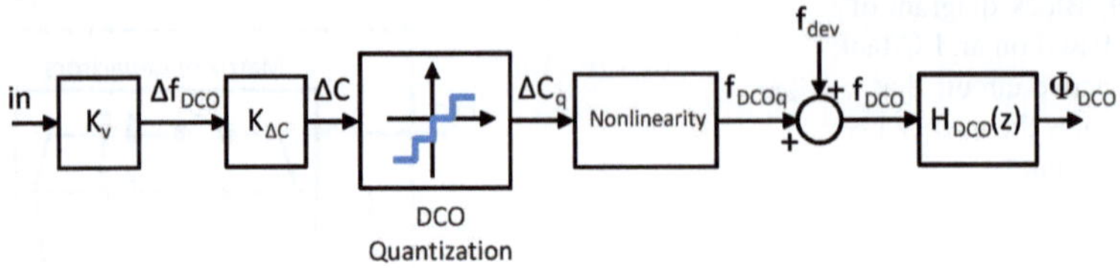

Fig. 4.10 Block diagram of a DCO with capacitive resolution res_{DCO} and additive frequency deviations f_{dev}

$$f_{DCO} = \frac{1}{\sqrt{L \cdot (C + \Delta C)}}. \tag{4.23}$$

By means of a Taylor series expansion we write from Eq. (4.18):

$$f_{DCO} \approx \frac{1}{\sqrt{LC}} + K_{\Delta C} \Delta C, \tag{4.24}$$

where $K_{\Delta C}$ is given by:

$$K_{\Delta C} = \frac{-L}{2\sqrt{(L\,C)^3}}. \tag{4.25}$$

Substituting Eq. (4.16) into Eq. (4.24) and solving for $K_{\Delta C}$, we obtain:

$$\Delta C = \frac{K_v in}{K_{\Delta C}}. \tag{4.26}$$

Equation (4.26) gives the variation in the capacitance of the LC-tank in a DCO with *infinite* resolution. The control word $ctrl$ in Fig. 4.9 can be related to $\Delta C / res_{DCO}$. For example, assume that the control word "ctrl" in Fig. 4.9 is equal the number of unit capacitors connected in parallel to the terminals of the matrix of capacitors. Hence, the control word $ctrl$ is equal to $\Delta C / res_{DCO}$.

In a DCO with *finite* resolution, the variation in the capacitance ΔC_q of the LC-tank is given by:

$$\Delta C_q = res_{DCO} \text{floor}\left(\frac{\Delta C}{res_{DCO}}\right), \tag{4.27}$$

where ΔC is given by Eq. (4.26).

The block diagram of a DCO with finite capacitive resolution res_{DCO} and additive frequency deviations f_{dev} is shown in Fig. 4.10. The output of the block labelled "DCO quantization" in Fig. 4.10 is given by Eq. (4.27). The output f_{DCOq} of the block labelled "Nonlinearity" in 4.10 is given by:

$$f_{DCOq} = \frac{1}{\sqrt{L \cdot (C + \Delta C_q)}}. \tag{4.28}$$

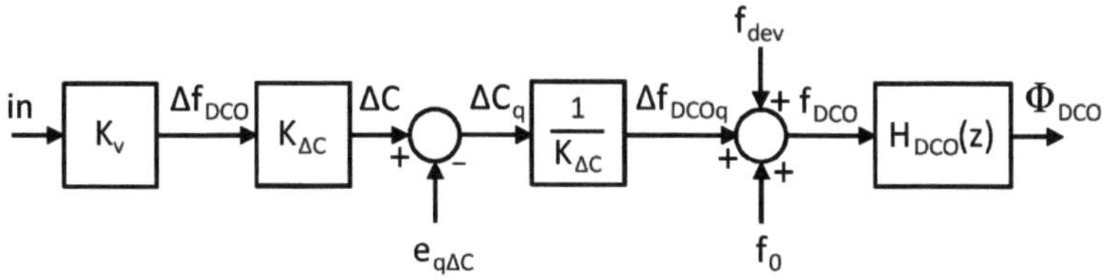

Fig. 4.11 Linearized model of a DCO with capacitive resolution res_{DCO} and additive frequency deviations f_{dev}

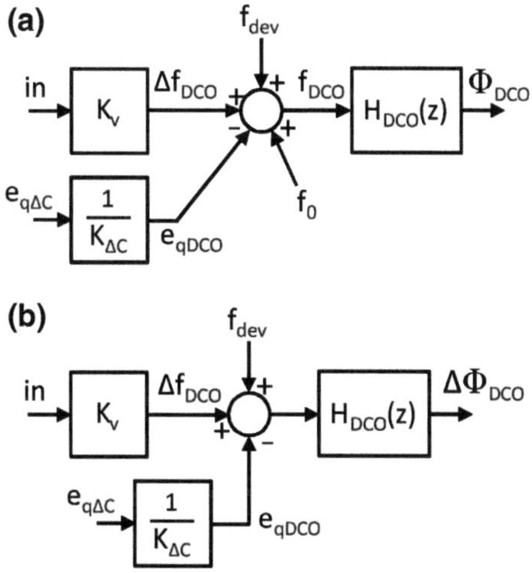

Fig. 4.12 DCO models **a** equivalent to that in Fig. 4.11 and **b** that relates the phase deviation $\Delta\Phi_{DCO}$ to the input in, the quantization error $e_{q\Delta C}$ and the additive frequency deviations f_{dev}

A linear model can be associated with the block diagram in Fig. 4.10. In fact, we can linearize Eq. (4.28) by means of a Taylor series expansion, as we have done for Eq. (4.23). Moreover, Eq. (4.27) can be represented by a linear additive noise model, as we have done for the quantizers in Sect. 4.3. By means of a Taylor series expansion and the additive noise model, we can obtain the model in Fig. 4.11 from that in Fig. 4.10.

The DCO quantization error $e_{q\Delta C}$ in Fig. 4.11 is given by:

$$e_{q\Delta C} = \Delta C - res_{DCO}\text{floor}\left(\frac{\Delta C}{res_{DCO}}\right). \tag{4.29}$$

The model in Fig. 4.11 can be simplified to that in Fig. 4.12a.

By setting $f_0 = 0$ in the model in Fig. 4.12a, we obtain the DCO model in Fig. 4.12b that relates the DCO phase deviation $\Delta\Phi_{DCO}$ to the input in, the quantization error $e_{q\Delta C}$ and the additive frequency deviations f_{dev}.

Finally, let us consider the case in which the matrix of capacitors in the DCO is driven by a sigma-delta modulator [10, 11]. A sigma-delta modulator can be used to increase the effective resolution of a DCO. Figure 4.13 shows an example block

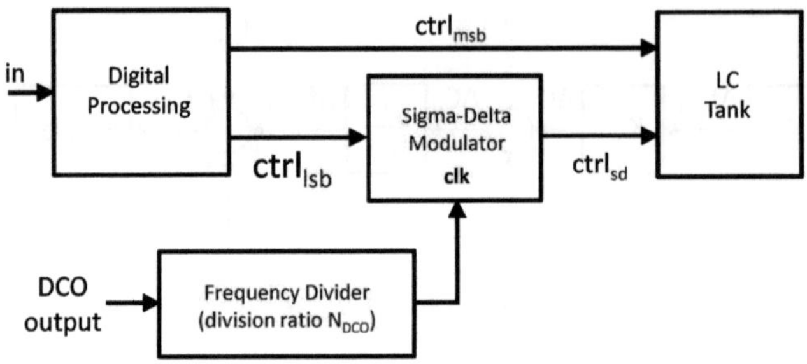

Fig. 4.13 Example block diagram of the circuits that produce the control signals of the matrix of capacitors in a DCO with a sigma-delta modulator

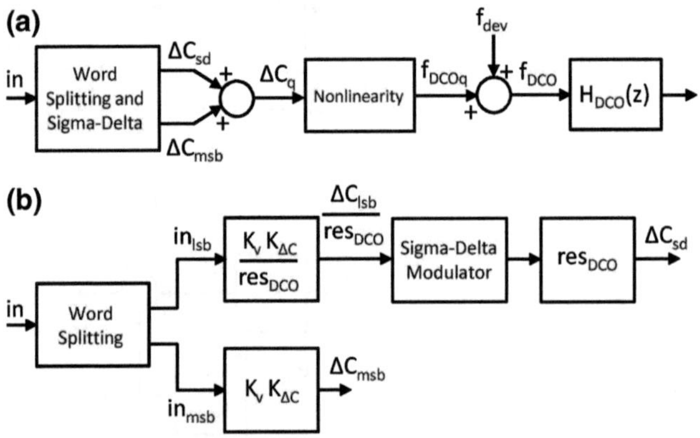

Fig. 4.14 Model associated with **a** the DCO with a sigma-delta modulator in Fig. 4.13 and **b** the block labelled "Word Splitting and Sigma-Delta" in (**a**)

diagram of a sigma-delta modulator used to drive the matrix of a capacitor in an LC DCO.

The block labelled "Digital Processing" in Fig. 4.13 splits the control word $ctrl$ of the LC tank in two parts $ctrl_{msb}$ and $ctrl_{lsb}$. Assume that $ctrl$, $ctrl_{msb}$ and $ctrl_{lsb}$ are fixed point digital words. The words $ctrl_{msb}$ and $ctrl_{lsb}$ are associated with the most and least significant bits, respectively, of the control word $ctrl$. For example, $ctrl_{lsb}$ and $ctrl_{msb}$ can be equal to the first three least significant bits and to the remaining bits in $ctrl$, respectively. In Fig. 4.13, the signal $ctrl_{lsb}$ is the input of a sigma-delta modulator which is clocked by a signal with a frequency equal that of the DCO divided by the division ratio N_{DCO}. The value of N_{DCO} determines the oversampling ratio of the sigma-delta modulator [12, 13]. The output $ctrl_{sd}$ of the sigma-delta modulator and $ctrl_{msb}$ determine the variations ΔC_{sd} and ΔC_{msb} in the capacitance of the LC tank. We assume that ΔC_{sd} and ΔC_{msb} are given by $res_{DCO} ctrl_{sd}$ and $res_{DCO} ctrl_{msb}$, respectively, where res_{DCO} is the capacitive resolution of the DCO. The total variation ΔC_q in the capacitance of the LC tank is given by the sum of ΔC_{sd} and ΔC_{msb}. The model in Fig. 4.10 can be modified in order to take into account ΔC_{sd} and ΔC_{msb}, as shown in Fig. 4.14a.

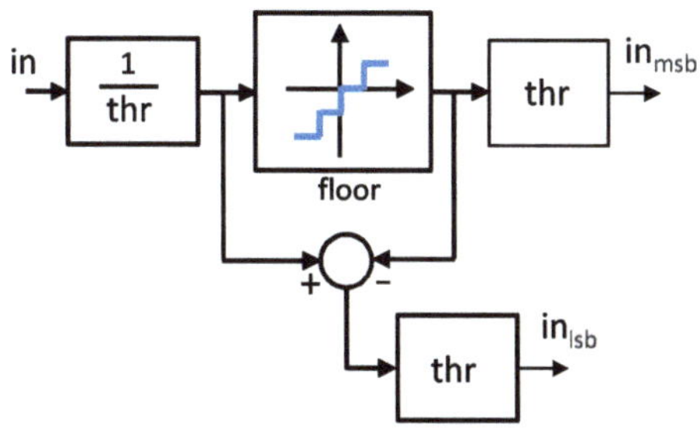

Fig. 4.15 Model of the block labelled "Word Splitting" in Fig. 4.14b

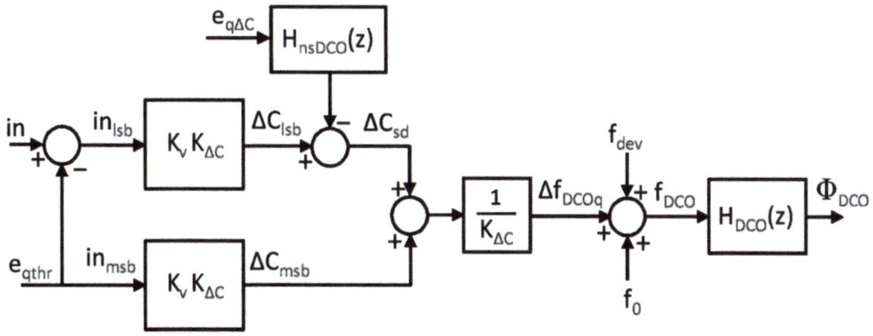

Fig. 4.16 Linear model of the DCO with a sigma-delta modulator in Fig. 4.13

The block labelled "Word Splitting and Sigma-Delta" in Fig. 4.14a is associated with the model in Fig. 4.14b. The block labelled "Sigma-Delta Modulator" in Fig. 4.14b represents the operations of the sigma-delta modulator in Fig. 4.13.

The block labelled "Word Splitting" in Fig. 4.14b is associated with the operation of the block labelled "Digital Processing" in Fig. 4.13. In fact, the signals $ctrl_{msb}$ and $ctrl_{lsb}$ can be related to in_{msb} and in_{lsb} by means of Eq. (4.24). From Eq. (4.24), we deduce that $ctrl_{msb}$ and $ctrl_{lsb}$ are equal to $K_v K_{\Delta C} in_{msb}/res_{DCO}$ and $K_v K_{\Delta C} in_{lsb}/res_{DCO}$, respectively. The splitting operation can be represented by a quantizer with threshold thr. A model of the block "Word Splitting" in Fig. 4.14b is shown in Fig. 4.15.

The block labelled "floor(·)" in Fig. 4.15 rounds to the closest integer smaller than the argument. Assume that nb_{lsb} is the number of bits of $ctrl$ that are associated with ctr_{lsb}, and that $ctrl$ is a fixed point word. The signal thr in Fig. 4.15 is the base-10 number equivalent of the binary number given by the register $ctrl$ with the first nb_{lsb} least significant bits and the remaining bits equal to one and zero, respectively.

The linear model in Fig. 4.16 can be obtained from the model in Fig. 4.14a by replacing the sigma-delta modulator and the quantizer in Figs. 4.14b and 4.15 with the additive noise model, respectively.

The block labelled "$H_{nsDCO}(z)$" in Fig. 4.16 represents the z-domain noise transfer function of the sigma-delta modulator in Fig. 4.14a. The transfer function $H_{nsDCO}(z)$ of an lth order sigma-delta modulator in the time domain of the edges of its clock is given by [3]:

$$H_{nsDCO}(z) = (1 - z^{-1})^l. \tag{4.30}$$

The time domain associated with Eq. (4.30) is defined by the period of the clock of the sigma-delta modulator that drives the DCO is equal to the DCO period T_{DCO} multiplied by the division ratio N_{DCO}. In order to write $H_{nsDCO}(z)$ in the discrete-time domain of the edges of the reference oscillator, we recall that $z = e^{sT}$, where T is the time interval between consecutive discrete-time events. By substituting $z = e^{s(N_{DCO}T_{DCO})}$ into Eq. (4.30), we obtain:

$$H_{nsDCO}(s) = (1 - z^{-sN_{DCO}T_{DCO}})^l. \tag{4.31}$$

The z variable associated with the time domain of the edges of the reference oscillator is related to the s-domain by $z = e^{sT_{Ref}}$. Moreover, we recall that $(e^a)^b = e^{ab}$ where a and b are real numbers. Thus, a transfer function that gives Eq. (4.31) when $z = e^{sT_{Ref}}$ is the following:

$$H_{nsDCO}(z) = \left(1 - z^{-\frac{N_{DCO}T_{DCO}}{T_{Ref}}}\right)^l. \tag{4.32}$$

Equation (4.32) represents the transfer function of the sigma-delta modulator that drives the DCO in the discrete-time domain of the edges of the reference oscillator.

The model in Fig. 4.16 can be simplified to that in Fig. 4.17a. By setting f_0 equal to zero in the model in Fig. 4.17a, it is possible to obtain the model that relates the DCO phase deviation $\Delta\Phi_{DCO}$ to the DCO input in, the DCO noise f_{dev} and the quantization error ΔC_q, shown in Fig. 4.17b.

4.5 Digital Filter

The digital filter in an ADPLL is represented by a difference equation and a z-domain transfer function in the time and frequency domains, respectively. We determine the transfer function of the digital filter in an ADPLL from a behavioral model of the ADPLL. In fact, we associate an ADPLL with an approximate s-domain transfer function at the behavioral level, as we will show in Sect. 4.6. We determine the coefficients of the s-domain ADPLL transfer function by means of equations that relate the s-domain transfer function to the ADPLL performance requirements in terms of step response, phase margin and bandwidth. We will discuss how to determine the coefficients in the ADPLL transfer function in Sect. 4.8.

We assume a digital filter associated with an s-domain transfer function $H_{LF}(s)$

$$H_{LF}(s) = K_{LF}\frac{p_1(s + z_1)}{(s + p_1)s}, \tag{4.33}$$

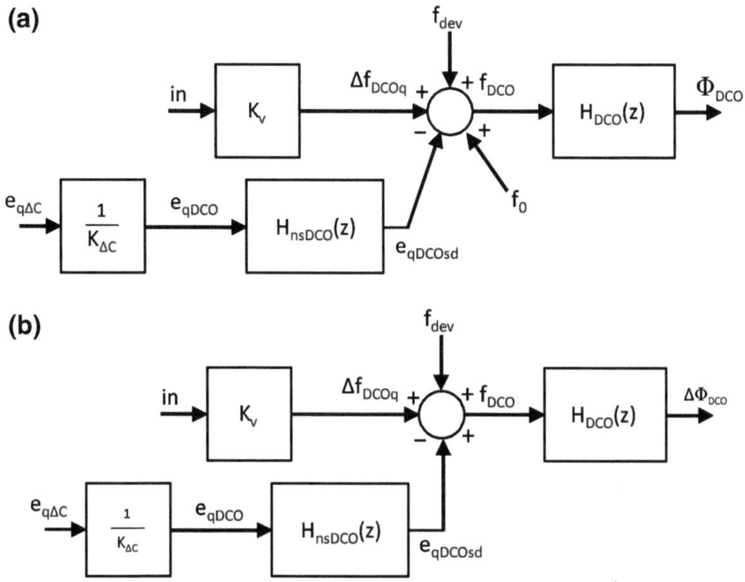

Fig. 4.17 **a** DCO model with sigma-delta modulator equivalent to that in Fig. 4.16 and **b** DCO model that relates the phase deviation $\Delta\Phi_{DCO}$ to the input in, the quantization error $e_{q\Delta C}$ and the additive frequency deviations f_{dev}

where the parameters K_{LF}, p_1 and z_1 determine the gain, pole and zero of the transfer function. Notice that $H_{LF}(s)$ includes a pole at the origin. Hence, an ADPLL with a digital filter associated with $H_{LF}(s)$ is a type-II ADPLL. Different from a type-I ADPLL, a type-II ADPLL does not require a trade-off between bandwidth and phase margin [14]. Moreover, different from a type-I ADPLL, a type-II ADPLL can drive the phase difference between the reference oscillator and the DCO to zero while having the output of the digital filter equal to a non-zero constant. Thus, a type-II ADPLL enables a more flexible design with respect to a type-I ADPLL.

By means of the bilinear transform we can write from Eq. (4.33):

$$H_{LF}(z) = K_{LF} \frac{(p_1 z + p_1)\left((z_1 + 2f_s)\,z + z_1 - 2f_s\right)}{(z_1 - 2f_s)\left((p_1 + 2f_s)\,z + p_1 - 2f_s\right)} \tag{4.34}$$

Applying the inverse z-transform to Eq. (4.34), we obtain a finite difference equation described by the block diagram in Fig. 4.18.

The outputs of the blocks labelled "Single Pole" and "Zero plus Integrator" in Fig. 4.18 are given by:

$$out[n] = a_1\, in[n] + a_2\, in[n-1] + b_1\, out[n-1] \tag{4.35}$$

and

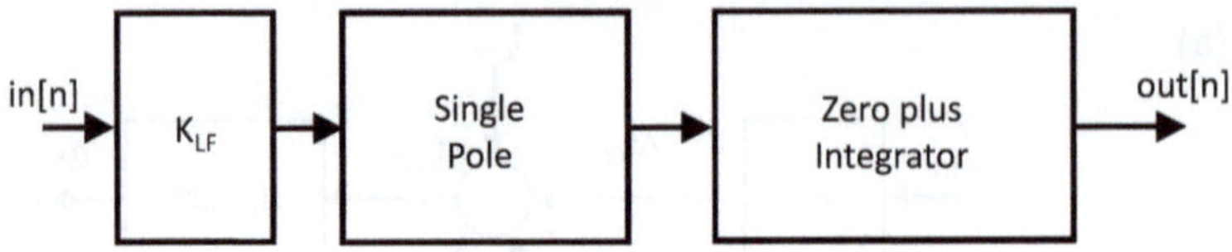

Fig. 4.18 Block diagram of the digital filter that we assume for an ADPLL

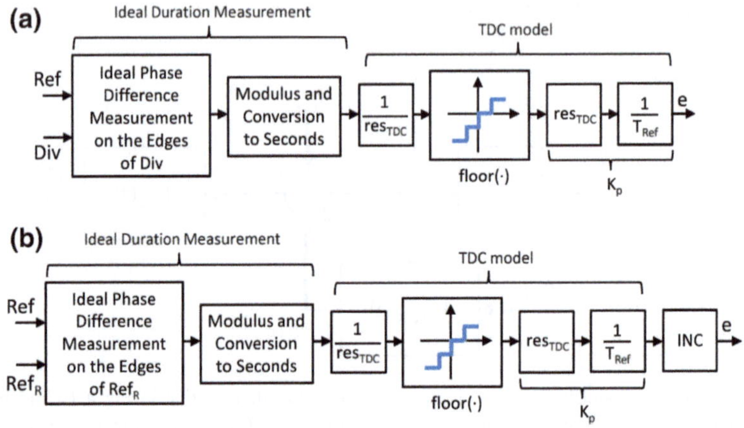

Fig. 4.19 Models representing the phase-to-digital conversion is **a** TDC-based and **b** accumulator-based ADPLLs when the integer part of the phase difference is equal to zero

$$out[n] = c_1\, in[n] + c_2\, in[n-1] + d_1\, out[n-1], \tag{4.36}$$

respectively. The coefficients a_1, a_2, b_1, c_1, c_2, d_1, in Eqs. (4.35) and (4.36) can be obtained by means of the bilinear transform from Eq. (4.34).

4.6 Full ADPLL Model

Figure 4.19a, b show the models that we derived in Chap. 2 in order to represent the phase-to-digital conversion in TDC-based and accumulator-based ADPLLs, respectively, when the integer part of the phase difference between the reference oscillator and the DCO is equal to zero.

The blocks labelled "TDC Model" in Fig. 4.19a, b are associated with zeroth-order noise-shaping TDCs. The model of lth-order noise-shaping TDCs can be included in Fig. 4.19a and b by replacing the blocks labelled "TDC Model" with the model in Fig. 4.7b, as shown in Fig. 4.20a, b, respectively.

In Fig. 4.20a, b, the blocks "Modulus and Conversion to Seconds" correspond to the blocks labelled "T_{Ref}" whose outputs are equal to their inputs multiplied by T_{Ref}. Moreover, the blocks "Ideal Phase Difference Measurement on the Edges of Div" and

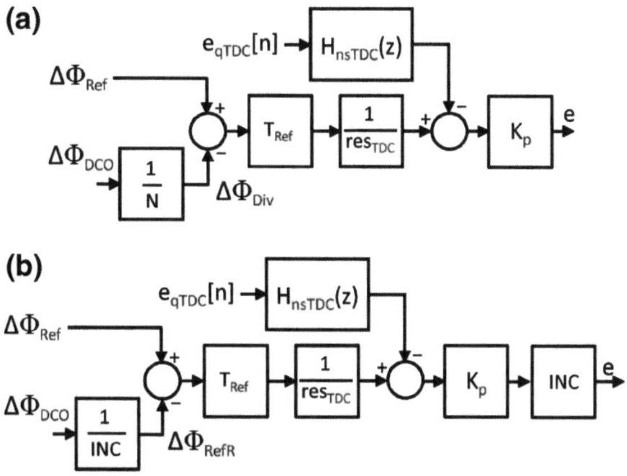

Fig. 4.20 Models representing the phase-to-digital conversion in **a** TDC-based and **b** accumulator-based ADPLLs with lth-order noise shaping TDCs when the integer part of the phase difference is equal to zero

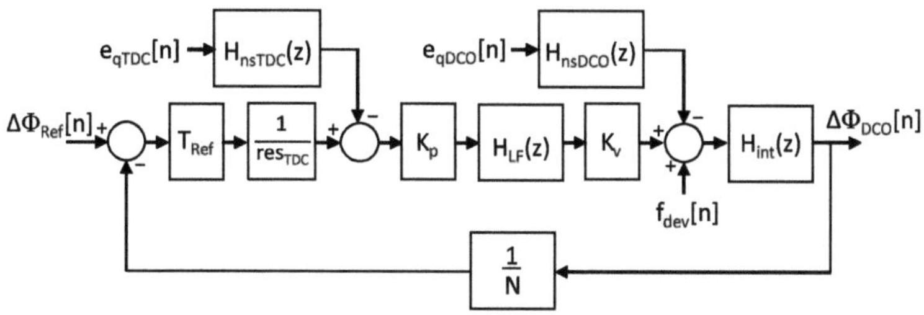

Fig. 4.21 Discrete time linear model of a TDC-based ADPLL with lth-order noise-shaping TDC and pth-order sigma-delta modulator driving the DCO

"Ideal Phase Difference Measurement on the Edges of Ref_R" are equivalent to the blocks that calculate $\Phi_{Ref} - \Phi_{Div}$ and $\Phi_{Ref} - \Phi_{RefR}$ in Fig. 4.20a, b, respectively. Notice that the phase Φ_{RefR} is equal to the DCO phase Φ_{DCO} divided by INC only when the integer part of the phase difference between Ref and Ref_R is equal to zero.

We obtain the linear model of a TDC-based ADPLL in Fig. 4.21 by replacing the blocks labelled "TDC", "Digital Filter", "DCO" and "Frequency Divider" in the block diagram of a TDC-based ADPLL in Fig. 4.1a by the models in Figs. 4.20a, 4.18 and 4.17, respectively.

The z-domain model in Fig. 4.21 is associated with the instants at which the rising edges of the reference oscillator occur. The transfer functions $H_{DCO}(z)$ and $H_{nsDCO}(z)$ are given by Eqs. (4.21) and (4.32), respectively.

The linear model of an accumulator-based ADPLL is shown in Fig. 4.22. We obtain the model in Fig. 4.22 by replacing the blocks labelled "Accumulator", "Counter", "TDC", "Digital Filter", "DCO" and "Frequency Divider" in the block

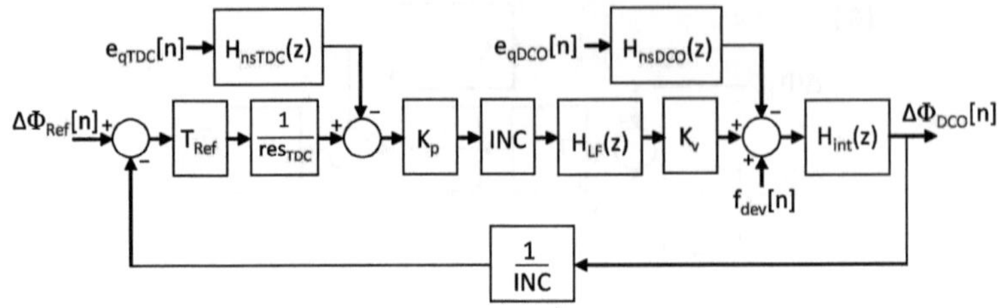

Fig. 4.22 Discrete time linear model of an accumulator-based ADPLL with lth-order noise-shaping TDC and pth-order sigma-delta modulator driving the DCO

diagram of an accumulator-based ADPLL in Fig. 4.1b by their linear models in Fig. 4.20b, 4.18 and 4.17, respectively.

The ADPLL models in Figs. 4.21 and 4.22 are linear and include three noise sources. Hence, the analytical prediction of the power spectral density $S_{DCO}(z)$ of the DCO phase deviation is given by:

$$S_{DCO}(z) = S_{DCO1}(z) + S_{DCO2}(z) + S_{DCO3}(z), \tag{4.37}$$

where $S_{DCO1}(z)$, $S_{DCO2}(z)$, $S_{DCO3}(z)$ are given by:

$$S_{DCO1}(z) = \left(\frac{res_{TDC}}{T_{Ref}}\right)^2 |H_{cl1}(z)H_{nsTDC}(z)|^2\, S_{qTDC}(z) \tag{4.38}$$

$$S_{DCO2}(z) = |H_{cl2}(z)|^2\, S_{fdev}(z), \tag{4.39}$$

and

$$S_{DCO3}(z) = |H_{cl2}(z)H_{nsDCO}(z)|^2\, S_{qDCO}(z). \tag{4.40}$$

In Eqs. (4.38), (4.39) and (4.40), $H_{cl1}(z)$ and $H_{cl2}(z)$ are given by:

$$H_{cl1}(z) = \frac{K_p K_v H_{LF}(z) H_{DCO}(z)}{1 + \frac{K_p K_v H_{LF}(z) H_{DCO}(z)}{N}}, \tag{4.41}$$

and

$$H_{cl2}(z) = \frac{H_{DCO}(z)}{1 + \frac{K_p K_v H_{LF}(z) H_{DCO}(z)}{N}}, \tag{4.42}$$

in the TDC-based ADPLL in Fig. 4.21. In the accumulator-based ADPLL in Fig. 4.22, $H_{cl1}(z)$ and $H_{cl2}(z)$ are given by:

$$H_{cl1}(z) = \frac{INC\, K_p K_v H_{LF}(z) H_{DCO}(z)}{1 + K_p K_v H_{LF}(z) H_{DCO}(z)}, \tag{4.43}$$

and

$$H_{cl2}(z) = \frac{H_{DCO}(z)}{1 + K_p K_v H_{LF}(z) H_{DCO}(z)}. \tag{4.44}$$

We assume that the power spectral densities $S_{qTDC}(z)$ and $S_{qDCO}(z)$ of the quantization errors in the TDC and the DCO are given by [3]:

$$S_{qTDC}(z) = \frac{(res_{TDC})^2}{12} \tag{4.45}$$

and

$$S_{qDCO}(z) = \frac{(res_{DCO})^2}{12}, \tag{4.46}$$

respectively. Equations (4.45) and (4.46) are based on the assumption that the quantization error in a quantizer is independent of the quantizer input. This assumption is not true in general. Hence, the accuracy of Eqs. (4.45) and (4.46) has to be verified by simulations or measurements.

Finally, we assume that the s-domain power spectral density $S_{DCO}(s)$ of the phase in a DCO with a constant input which is *not included* in an ADPLL is given by:

$$S_{DCO}(s) = \frac{var_{fdev}}{|s|^2}, \tag{4.47}$$

where var_{fdev} determines the power of the DCO phase noise. The phase noise model associated with Eq. (4.47) is common in the literature [2, 6] and is consistent with the phase noise model used in CppSim [7]. By applying the bilinear-transform to Eq. (4.47), we can write:

$$S_{DCO}(z) = var_{fdev} \left| \frac{T_{Ref}\,(z+1)}{2(z-1)} \right|^2 \tag{4.48}$$

where T_{Ref} is the period of the reference oscillator and $S_{DCO}(z)$ is the z-domain power spectral density of the phase in a DCO with a constant input which is not included in an ADPLL. Notice that Eq. (4.48) gives the power spectral density at the output of an *ideal integrator* when a white noise signal with variance var_{fdev} is applied at the integrator input. Hence, the PSD of the DCO phase deviation is given by Eq. (4.48) when the signal f_{dev} in the DCO model in Fig. 4.17 is white noise with variance var_{fdev} given by:

$$var_{fdev} = \frac{\Delta f^2 \cdot 10^{\frac{S_{DCO}(\Delta f)}{10}}}{T_{Ref}}, \tag{4.49}$$

where $S_{DCO}(\Delta f)$ is the power spectral density in dBc/Hz at the frequency offset Δf. Equation (4.49) is used by CppSim in the model of a VCO with phase noise.

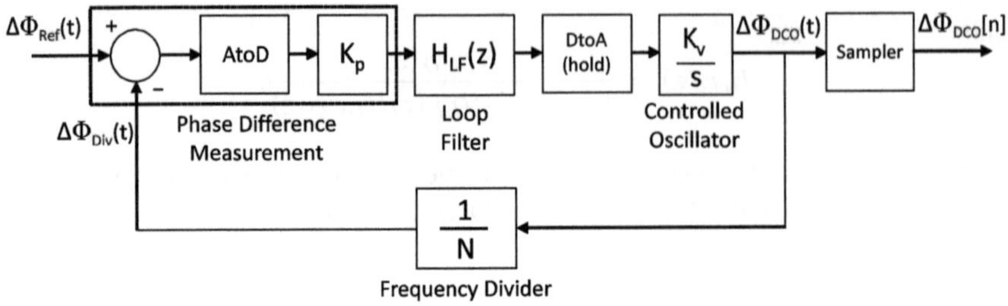

Fig. 4.23 Linear mixed between s- and z-domain model of an ADPLL that relates the phase deviations of the reference oscillator to those of the DCO

4.7 Mixed-Signal Models

The scripts that we report in this chapter are based on finite difference equations that are associated with a z-domain ADPLL model. It is important to emphasize that an ADPLL s-domain model does not include the effects of the analog-to-digital and digital-to-analog conversions that are implemented by the TDC and the DCO, respectively. In order to model these mixed-signal effects, we consider a mixed s- and z-domain ADPLL model, as shown in Fig. 4.23.

The blocks labelled "AtoD" and "DtoA (hold)" in Fig. 4.23 represent the analog-to-digital and digital-to-analog conversions that are implemented by the TDC and the DCO, respectively. The digital filter in the ADPLL model in Fig. 4.23 is defined by a discrete-time transfer function $H_{LF}(z)$.

Assume that the phase difference calculation is realized by measuring the time interval between an edge of the reference oscillator and the next edge of the frequency divider. Hence, the time interval that separates consecutive phase difference measurements is at least one period of the reference oscillator. As a consequence, it is possible to assume that the period of the clock of the digital filter is very close to that of the reference oscillator. Thus, we can assume that the hold time of the "DtoA (hold)" block in Fig. 4.23 is approximatively equal to the period of the reference oscillator.

The model in Fig. 4.23 can be used to predict the spectrum of the phase deviations of the DCO. We recall that sampling a signal at frequency F_{Ref} produces spectral replicas of the spectrum of the signal shifted by integer multiples of F_{Ref} in the frequency domain [5]. Furthermore, when the samples of a signal are held for time intervals with durations $1/F_{Ref}$, the spectrum of the "sampled and held" signal is equal to the spectrum of the sampled signal (which includes the replicas) multiplied by a $sinc$ function with zeros at multiples of F_{Ref} in the frequency domain [5]. In order to simplify the analysis, we assume that the bandwidth of the ADPLL is much smaller than F_{Ref} and that the spectral replicas can be neglected because they are attenuated by the $sinc$ function that is associated with the "holding" operation. When the ADPLL bandwidth is much smaller than F_{Ref}, the transfer function $H_{LF}(z)$ can be replaced by its s-domain equivalent that is given by the bilinear transform. We

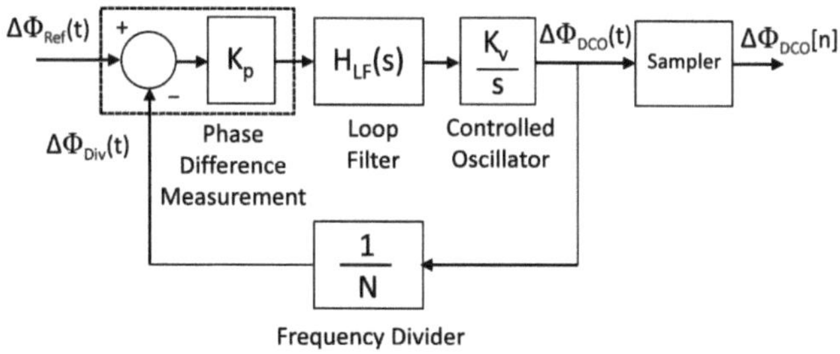

Fig. 4.24 Linear s-domain model of an ADPLL that relates the phase deviations of the reference oscillator to those of the DCO

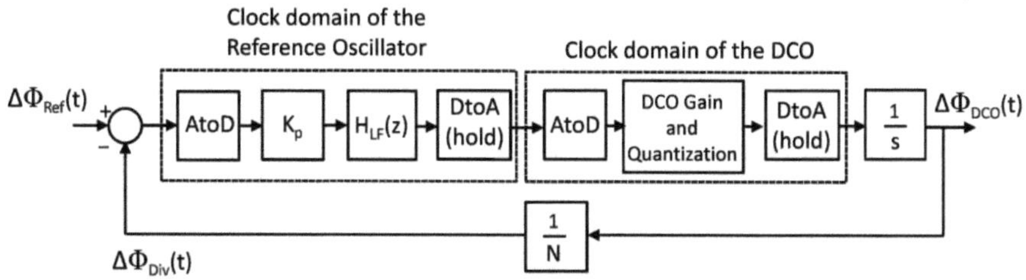

Fig. 4.25 Example phase deviation $\Delta\Phi(t)$, ideal and perturbed phases of an oscillator

conclude that the model in Fig. 4.23 approximates that in Fig. 4.24 for frequencies that are much smaller than the ADPLL bandwidth. Notice that the s-domain model in Fig. 4.24 can be related to the z-domain model in Fig. 4.21 by means of the bilinear transform.

The model in Fig. 4.23 can be simulated using Verilog-AMS or Simulink. For example, the modelling technique that we will present in Chap. 6 can be used to simulate in Simulink mixed models such as that in Fig. 4.23. However, when the ADPLL bandwidth is much smaller than F_{Ref}, i.e., when the model in Fig. 4.23 is equivalent to those in Figs. 4.21 and 4.24, we can use s- or z-domain models to simulate the phase noise performance of an ADPLL.

Finally, we note that the model in Fig. 4.23 can be modified to take account of the operations of the ADPLL that happen in the time domain of the DCO edges. For example, a sigma-delta modulator that is clocked by the edges of the DCO is sometimes used to control the DCO [10, 11], as we have discussed in Sect. 4.4. For further analysis of an ADPLL with a DCO driven by a sigma-delta modulator, we refer the reader to [10, 12].

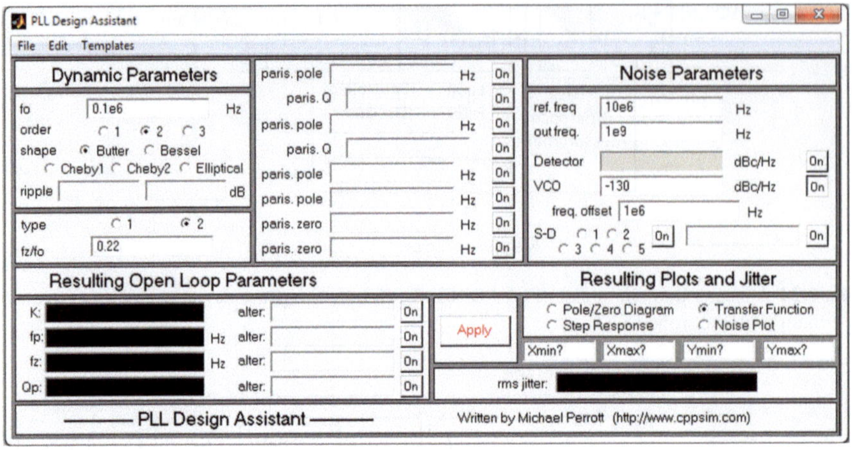

Fig. 4.26 Reference parameters in the PLL design assistant

4.8 An Example TDC-Based ADPLL

We use the PLL Design Assistant [15] to determine the design parameters of an example TDC-based ADPLL. The PLL Design Assistant designs traditional PLLs based on PFDs, charge pumps and VCOs. Nevertheless, the z-domain transfer function of an ADPLL can be related to the equivalent s-domain transfer function that is designed by the PLL Design Assistant, as shown in the previous sections. The design parameters given by the PLL Design Assistant can be used for an ADPLL by assuming that the gains of the PFD and the VCO in the PLL Design Assistant are equivalent to those of a TDC and a DCO in an ADPLL, respectively. Moreover, the coefficients of the s-domain analog filter given by the PLL Design Assistant are related to those of an equivalent z-domain digital filter by the bilinear transform. The graphical interface of the PLL Design Assistant is shown in Fig. 4.26.

The example TDC-based ADPLL has the frequency of the reference oscillator and the free-running DCO frequency equal to 10 MHz and 1 GHz, respectively, as shown in Fig. 4.26. The phase noise of the DCO is equal to −130 dBc/Hz when the frequency offset is 1 MHz. The design parameters are related to a type-II ADPLL with a second order loop filter. The transfer function of the simulated ADPLL is third order. Figure 4.27 shows the phase noise of the VCO predicted by the PLL Design Assistant for a PLL associated with the design parameters in Fig. 4.26.

The VCO phase noise in Fig. 4.27 can be seen as the lower limit for an ADPLL based on a TDC and a DCO, each of which has infinite resolution.

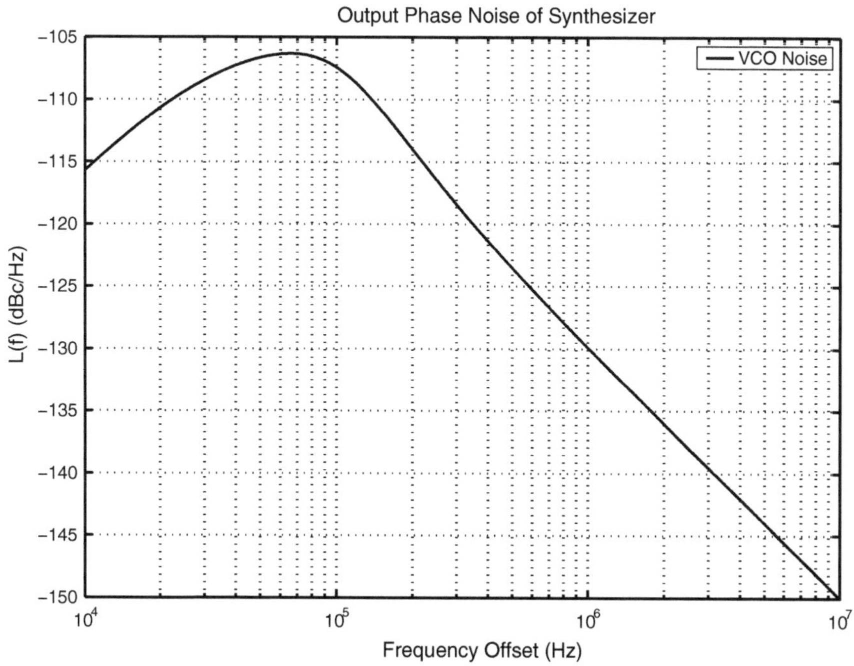

Fig. 4.27 Phase noise predicted by the PLL Design Assistant for an ADPLL associated with the design parameters in Fig. 4.26

4.9 Simulations with Machine-Precision TDC and DCO

The model of a TDC-based ADPLL with a TDC and a DCO, both of which have machine precision resolution, is implemented by the Matlab script in List. 8.1 that can be found in Appendix 8.1. The script in List. 8.1 is based on the ADPLL models described in the previous sections with the design parameters shown in Fig. 4.26.

Figure 4.28 shows the phase noise of the DCO simulated with the Matlab model in List. 8.1 and predicted by Eq. (4.37). Since the DCO and TDC have time and capacitive resolution equal to the machine precision, the noise components S_{DCO1} and S_{DCO3} in Eq. (4.37) are negligible. It is important to notice that the analytical predictions in Fig. 4.28 are s-domain transfer functions obtained from Eq. (4.37) by means of the bilinear transform in order to match the s-domain predictions of the PLL Design Assistant. The analytical predictions and the simulations match very well, except at high frequencies. The analytical predictions are inaccurate at high frequencies because of numerical errors in the simulation.

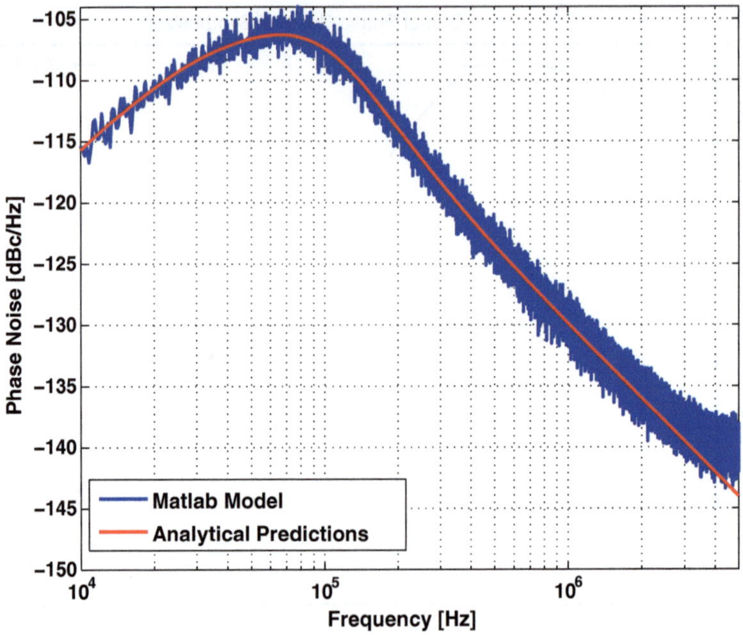

Fig. 4.28 Phase noise of an ADPLL which includes a TDC and a DCO with machine precision resolution from Matlab simulations and analytical predictions

4.10 Simulations with a Finite-Resolution TDC

The quantization in the TDC significantly degrades the phase noise performance of an ADPLL. In order to show the effect of the quantization in the TDC on the phase noise performance of an ADPLL, we consider the ADPLL model in List. 8.2 that is reported in Appendix 8.1. The ADPLL in List. 8.2 includes a TDC with finite time resolution equal to 10 ps and it has the same design parameters and the same DCO noise as the ADPLL in List. 8.1. Furthermore, the model in List. 8.2 includes the possibility to add dither to the TDC input. Dither is added directly to the duration of the time interval that the TDC has to convert to a digital word, as shown in Fig. 4.29.

We first consider the case when dither is *not* added to the TDC input. Then, we will show simulation results when the TDC input is dithered.

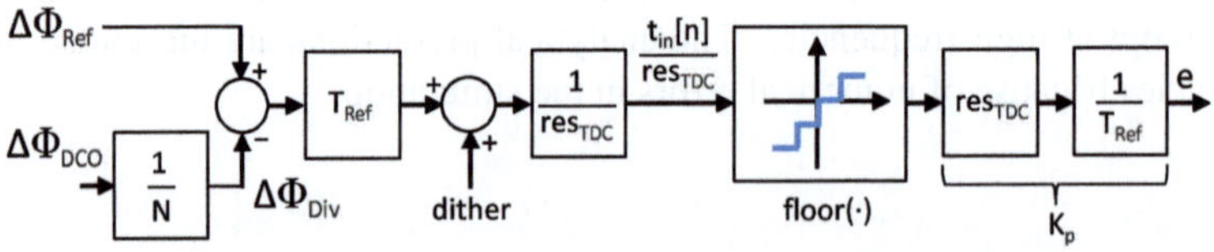

Fig. 4.29 Block diagram that shows how dither is added to the TDC input in the ADPLL model

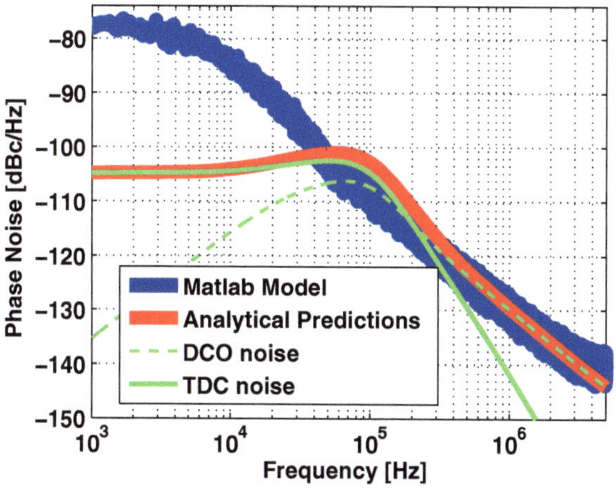

Fig. 4.30 Analytical predictions and simulations of the phase noise of an ADPLL which includes a TDC with finite resolution and a DCO with machine precision resolution. Dither is *not* added to the TDC input

Fig. 4.31 PSD of the TDC quantization error in the ADPLL model associated with Fig. 4.30

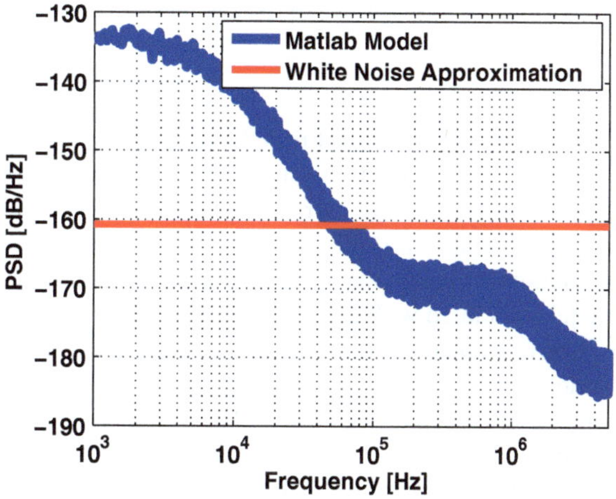

Figure 4.30 shows analytical predictions and simulations of the phase noise of an ADPLL model in List. 8.2 when dither is *not* added to the TDC input. The dither added to the TDC input in List. 8.2 is equal to zero when the variable `swichTDCdtr` is set to zero.

Simulations and predictions do not match in Fig. 4.30. Moreover, the phase noise at low frequencies is increased significantly with respect to Fig. 4.28. Figure 4.31 shows the PSD of the TDC quantization error in the model in List. 8.2. The quantization error associated with the PSD in Fig. 4.33 is expressed in cycles of the reference oscillator.

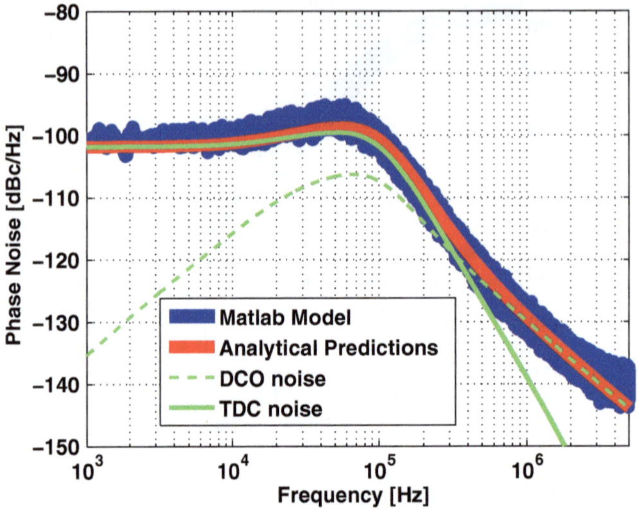

Fig. 4.32 Analytical predictions and simulations of the phase noise of an ADPLL which includes a TDC with finite resolution and a DCO with machine precision resolution. Dither is added to the TDC input

The horizontal line in Fig. 4.30 is the analytical prediction of the TDC quantization error obtained by means of the white noise approximation. The white noise approximation is *not accurate* for the PSD of the quantization error in the ADPLL model associated with Fig. 4.30, as shown in Fig. 4.31. We conclude that analytical predictions based on the white noise approximation are not accurate in the ADPLL model associated with Fig. 4.30 when dither is not added to the TDC input.

Let us consider the case when dither is added to the TDC input in ADPLL model in List. 8.2. Dither is added to the TDC input in List. 8.2 when the variable `swichTDCdtr` is set to 1. The dither added to the TDC input is white with a variance equal to the variance of the TDC quantization error. Figure 4.32 shows the simulated phase noise ADPLL model in List. 8.2 with dither.

The curve labelled "TDC noise" in Fig. 4.32 takes into account the TDC quantization error *and* the dither. The quantization in the TDC and the dither degrade the phase noise at low frequencies, as can be seen by comparing Figs. 4.28 and 4.32. The analytical predictions and simulations in Fig. 4.32 match well.

Figure 4.33 shows the PSD of the TDC quantization error in the model in the script in List. 8.2.

The white noise approximation is accurate when dither is added to the TDC input, as shown in Fig. 4.33. We conclude that it is necessary to add dither to the TDC input in order to make the white noise approximation accurate.

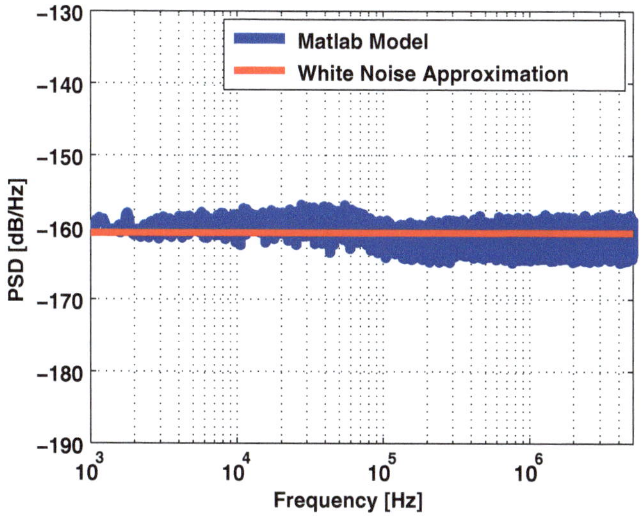

Fig. 4.33 PSD of the TDC quantization error in the ADPLL model associated with Fig. 4.32

4.11 Simulations with a Noise-Shaping TDC

The phase noise performance of an ADPLL at low frequencies is improved by a first-order noise-shaping TDC. Figure 4.34 shows the simulated and analytically predicted phase noise performance of an ADPLL which includes a first-order noise-shaping TDC with the same nominal resolution as the TDC in the ADPLL that is associated with Figs. 4.30 and 4.32. It is important to notice that dither has *not* been added to the input of the first-order noise-shaping TDC. The phase noise performance at low frequency in the ADPLL with the first-order noise-shaping TDC is limited by the DCO noise. Furthermore, the phase noise performance in Fig. 4.34 is very close to that of an ADPLL without quantization in Fig. 4.28. Figure 4.34 is obtained with the Matlab script in List. 8.3 that is reported in Appendix 8.1.

The maximum value of the analytical predictions of the TDC phase noise in Fig. 4.34 is equal to -130 dBc/Hz at a frequency offset equal to 10^5 Hz, approximately. By contrast, the maximum value of the analytical predictions of the TDC phase noise in Fig. 4.32 is equal to -106 dBc/Hz at a frequency offset equal to 10^5 Hz, approximately, which is 24 dBc/Hz above the maximum in Fig. 4.34. The difference between the maximum value of the analytical predictions of the TDC phase noise in Figs. 4.32 and 4.34 is related to the squared modulus of the first-order shaped transfer function H_{nsTDC} given by Eq. (4.15) in the noise-shaping TDC. In fact, the squared modulus of H_{nsTDC} is 24 dB when T_s and Ω are equal to $T_{Ref} = 0.1\,\mu$s and $\Omega = 2 * pi * 10^5$.

Figure 4.35 shows the PSD of the quantization error in the first-order noise-shaping TDC that is associated with Fig. 4.34. The analytical prediction based on the white

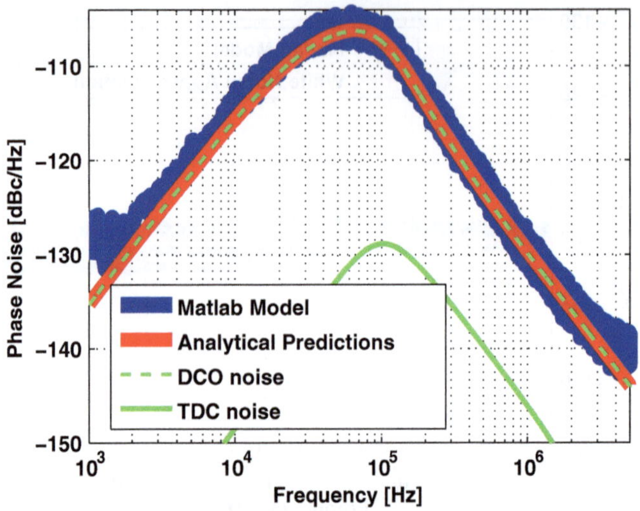

Fig. 4.34 Analytical predictions and simulations of the phase noise of an ADPLL which includes a *first-order noise-shaping* TDC with finite resolution and a DCO with machine precision resolution. Dither is *not* added to the TDC input

noise approximation is not very accurate in the ADPLL with first-order noise-shaping TDC and no dither, as shown in Fig. 4.35.

We conclude that introducing a first-order noise-shaping TDC in an ADPLL can improve the noise performance at low frequency significantly.

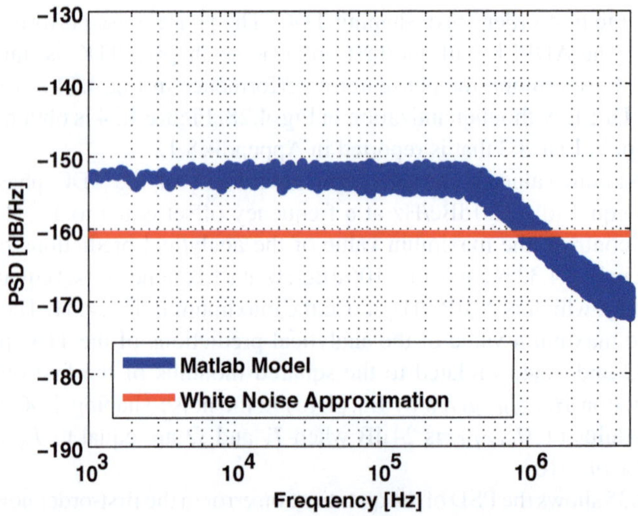

Fig. 4.35 PSD of the TDC quantization error in the ADPLL model associated with Fig. 4.34

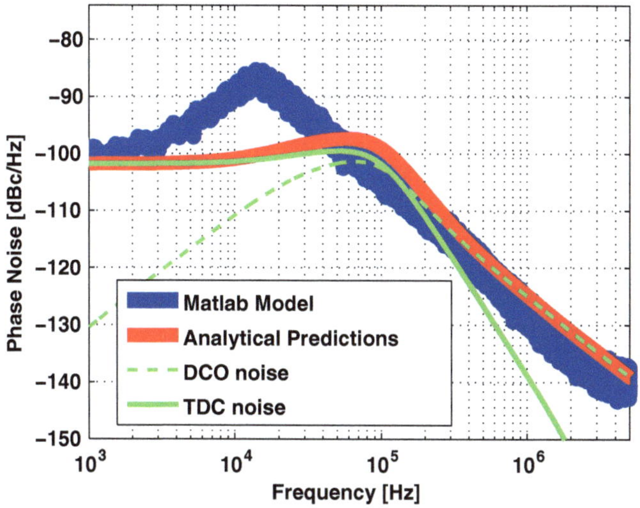

Fig. 4.36 Analytical predictions and simulations of the phase noise of an ADPLL which includes a TDC and a DCO with finite resolutions equal to 10 ps and 0.1 fF, respectively. Dither is added to the TDC input. The DCO input is *not* dithered

4.12 Simulations with a Finite-Resolution TDC and a Finite-Resolution DCO

A DCO with finite resolution degrades the noise performance of an ADPLL that includes a finite-resolution TDC with additive input dither. Figure 4.36 shows the simulated and predicted phase noise of the ADPLL in List. 8.2 that is reported in Appendix 8.1 when the resolutions of the TDC and the DCO are set to 10 ps and 0.1 fF, respectively.

Notice that it is possible to add dither to the DCO input in the script in List. 8.2 by setting the variable `swichDCOdtr` to 1. Dither is added to the input control word of the DCO, directly. Figure 4.36 has been produced with the variable `swichDCOdtr` set to 0, and dither is *not* added to the DCO input. In Fig. 4.36 dither is added to the TDC input and the resolution of the TDC is the same as that in Fig. 4.33. The phase noise at low frequencies in Fig. 4.36 is higher than that in Fig. 4.33. The dashed line in Fig. 4.36 is the analytical prediction that is associated with the noise *and* quantization error in the DCO. The analytical predictions do not match well the simulations in Fig. 4.36.

Figure 4.37 shows the simulated and predicted phase noise of the ADPLL in List. 8.2 when the DCO resolution is equal to 0.1 fF and dither is *added* to the DCO input.

The dither added to the DCO input is represented by integer numbers uniformly distributed over the interval [−4, 3]. The phase noise at low frequencies in Fig. 4.37 is lower than that in Fig. 4.36. Analytical predictions and simulations match well in

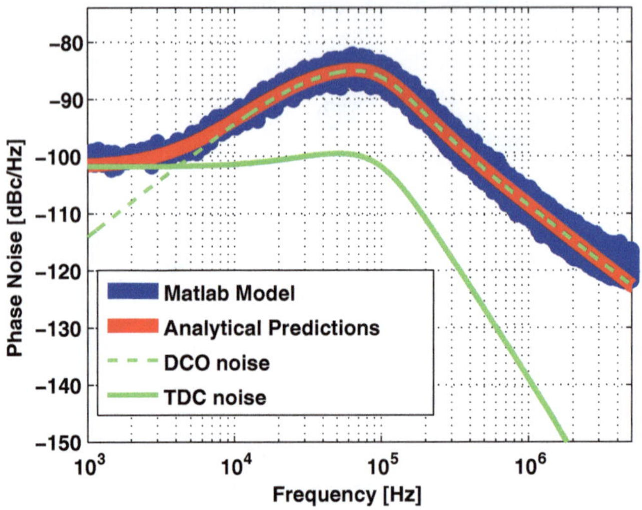

Fig. 4.37 Analytical predictions and simulations of the phase noise of an ADPLL which includes a TDC and a DCO with finite resolutions equal to 10 ps and 0.1 fF, respectively. The inputs of the TDC and the DCO are dithered

Fig. 4.37. We conclude that dither has to be added to the DCO input when the DCO has a finite resolution in order to reduce the phase noise at low frequencies and to increase the accuracy of analytical predictions based on the white noise approximation.

4.13 Conclusion

In this chapter, we have derived analytical equations to predict the phase noise in TDC-based and accumulator-based ADPLLs. The analytical equations that were derived are based on linear models. In fact, we have first shown linear models for the building blocks of TDC-based and accumulator-based ADPLLs. Then we have related the full ADPLL architecture to a linear model and to analytical equations for the phase noise. We have also reported Matlab scripts to simulate an ADPLL and have compared simulated and analytically estimated power spectral densities. When a TDC is used, it should be dithered. The DCO has to be dithered. Noise shaping in the TDC improves the phase noise performance significantly.

References

1. A. Abidi, "Basics of Phase Noise and Jitter", IEEE SSCS Chapter, Apr. 2011. Available at: http://ewh.ieee.org/r7/toronto/chapters/ssc/JitterXandXPhaseXNoise.pdf
2. K. Kundert, "Predicting the Phase Noise and Jitter of PLL-Based Frequency Synthesizers", Designer's Guide Consulting Inc., downloadable at http://www.designers-guide.org/analysis/PLLnoise+jitter.pdf. Last access in October 2012

3. R. Schreier, G. C. Temes, "Understanding Delta-Sigma Data Converters", IEEE Press, John Wiley & Sons Inc., 2005.
4. B. Widrow, I. Kollar, "Quantization Noise", Cambridge University Press, 2008.
5. S. Haykin, B. Van Veen, "Signals and Systems", Second Edition, John Wiley & Sons Inc., 2003.
6. M. Perrott, "Tutorial on Digital Phase-Locked Loops CICC 2009", Sep. 2009 Downloadable at: http://www.cppsim.com/PLL_Lectures/digital_pll_cicc_tutorial_perrott.pdf Accessed in Sep. 2013
7. M. Perrott, "CppSim Reference Manual Version 5", Aug. 2008 Downloadable at: http://www.cppsim.com/Manuals/cppsimdoc.pdf Accessed in Sep. 2013
8. A. V. Oppenheim, R. V. Schafer, "Discrete-Time Signal Processing", Third Edition, Pearson Higher Education Inc., 2010.
9. R. B. Staszewski, P. T. Balsara, "All-Digital Frequency Synthesizer in Deep-Submicron CMOS", John Wiley & Sons Inc., 2006.
10. R. B. Staszewski, K. Waheed, S. Vemulapalli, F. Dulger, J. Wallberg, C.-M. Hung, O. Eliezer "Spur-Free All-Digital PLL in 65nm for Mobile Phones" Solid-State Circuits Conference Digest of Technical Papers (ISSCC 2011), pp. 52–54, Feb. 2011.
11. N. Pavlovic, J. Bergervoet, "A 5.3GHz Digital-to-Time-Converter-Based Fractional-N All-Digital PLL' Solid-State Circuits Conference Digest of Technical Papers (ISSCC 2011), pp. 54–56, Feb. 2011.
12. R. B. Staszewski, C.-M. Hung, D. Leipold and P. T. Balsara, "A first multigigaherts digitally controlled oscillator for wireless applications", IEEE Trans. on Microwave Theory and, Techniques, 51(11), pp. 2154–2164, Nov. 2003.
13. R. B. Staszewski, C. Fernando, and P. T. Balsara, "Event-driven simulation and modeling of phase noise of an RF oscillator", IEEE Trans. Circuits Sys. I, vol. 52, no. 4, pp. 723–733, Apr. 2005.
14. B. Razavi, "RF Microelectronics", Pearson Education Inc., Second Edition, 2012.
15. M. Perrott, "PLL Design Using the PLL Design Assistant Program", Apr. 2005 Downloadable at: http://www.cppsim.com/Manuals/pll_manual.pdf Accessed in Sep. 2013

Chapter 5
Advantages of Noise Shaping and Dither

5.1 Introduction

We have shown in Chap. 3 that zeroth and first-order noise-shaping TDCs can be modelled by quantizers and first-order sigma-delta modulators, respectively. In this chapter, we consider the cases in which a dither signal $dtr[n]$ is added to the input $in[n]$ of a quantizer or a sigma-delta modulator followed by a moving average filter. In order to keep our analysis simple, we assume that $dtr[n]$ is uniformly distributed over the interval $[-\Delta d/2, \Delta d/2]$. Dither has the effect of making the quantization errors that are associated with the quantizer and the sigma-delta modulator more white and uniformly distributed over their intervals of definition. The moving average filters remove part of the noise associated with the dither and the quantization error. The removal of part of the power of the quantization error corresponds to an increase in the effective precision of a quantizer or a sigma-delta modulator when followed by a moving average filter. We determine analytically the precisions of these systems in terms of the maximum difference between the input and the output when this difference is bounded. In the cases where the maximum difference between the input and output is unbounded (because its distribution is Gaussian, for example), we assume that a measure of the precision of a quantizer or a sigma–delta modulator is the size of the interval $[-3\sigma, 3\sigma]$ of the contributions of the quantization error and the dither to the output, where σ is the standard deviation.

5.2 Quantizers and Sigma-Delta Modulators with Dither

In this section, we will determine the difference between the inputs and outputs of the systems "Quantizer plus Moving Average Filter" and "Sigma-delta Modulator plus Moving Average Filter" that are shown in Fig. 5.1a and b, respectively. Dither $dtr[n]$ is added to the input $in[n]$ of each system. In order to keep our analysis simple, we assume that $dtr[n]$ is white and uniformly distributed over the interval

F. Brandonisio and M. P. Kennedy, *Noise-Shaping All-Digital Phase-Locked Loops*, Analog Circuits and Signal Processing, DOI: 10.1007/978-3-319-03659-5_5,

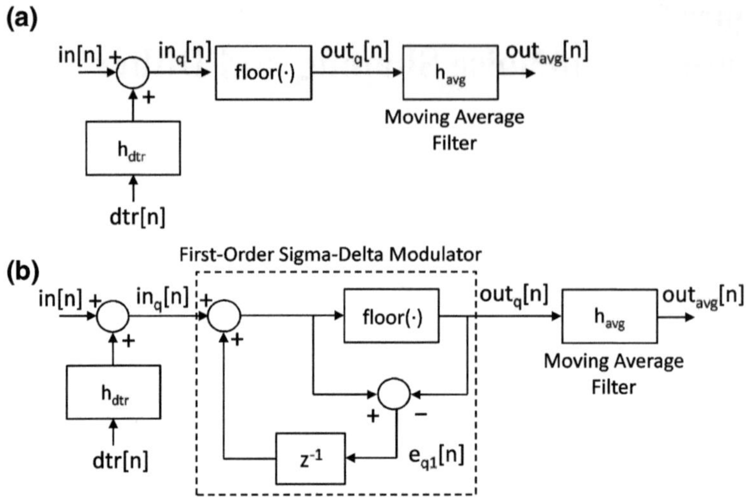

Fig. 5.1 Block diagrams of **a** a "quantizer plus moving average filter" and **b** a "sigma-delta modulator plus moving average filter" with dither

$[-\Delta d/2, \Delta d/2]$. The block labelled "$h_{(dtr)}$" represents a time domain operation that is defined by the time-discrete impulse response $h_{(dtr)}[n]$ that can be used to change the power spectral density of $dtr[n]$. We consider two example cases of $h_{(dtr)}[n]$:

$$h_{dtr0}[n] = \delta[n] \tag{5.1}$$

and

$$h_{dtr1}[n] = \delta[n-1] - \delta[n], \tag{5.2}$$

where $\delta[n]$ is an ideal discrete-time domain impulse. Equations (5.1) and (5.2) give zeroth- and first-order shaped dithers, respectively.

The impulse response $h_{avg}[n]$ of the moving-average filters in Fig. 5.1a, b is given by:

$$h_{avg}[n] = \frac{1}{N_{avg}} \sum_{i=0}^{N_{avg}-1} \delta[n-i], \tag{5.3}$$

where in_{filt} is the input of the moving-average filter and N_{avg} is the number of samples that are used by the filter. A moving-average filter is often referred to as a "sinc" filter because the Discrete Time Fourier Transform of its impulse response is a sinc function [1].

The output $out_q[n]$ of the quantizer that is represented by the floor(·) block in Fig. 5.1a is given by:

$$out_q[n] = \text{floor}(in_q[n]), \tag{5.4}$$

where in_q is the input of the quantizer. The quantization error $e_{q0}[n]$ of a quantizer with unity gain is defined by:

$$e_{q0}[n] = in_q[n] - out_q[n]. \tag{5.5}$$

The output $out_q[n]$ of the first-order sigma-delta modulator that is shown in Fig. 5.1b is defined by:

$$out_q[n] = \text{floor}(in_q[n] + e_{q1}[n-1]), \tag{5.6}$$

where the quantization error $e_{q1}[n]$ is given by:

$$e_{q1}[n] = in_q[n] + e_{q1}[n-1] - out_q[n]. \tag{5.7}$$

Equations (5.5) and (5.7) can be rewritten as follows:

$$out_q[n] = in_q[n] - e_{q0}[n] \tag{5.8}$$

and

$$out_q[n] = in_q[n] - e_{q1}[n] + e_{q1}[n-1], \tag{5.9}$$

respectively. Equations (5.8) and (5.9) can be written as a single expression:

$$out_q[n] = in_q[n] - H_{ns}\{e_q[n]\}, \tag{5.10}$$

where $H_{ns}\{\cdot\}$ is the operator whose impulse response is $h_{ns}[n]$. $h_{ns}[n]$ is given by:

$$h_{ns0}[n] = \delta[n], \tag{5.11}$$

or

$$h_{ns1}[n] = \delta[n-1] - \delta[n], \tag{5.12}$$

in a quantizer or first-order sigma-delta modulator, respectively. Equation (5.10) is represented by the block diagram in Fig. 5.2.

By replacing the "floor(·)" block and the sigma delta modulator in Fig. 5.1a and b, respectively, by the block diagram in Fig. 5.2, we obtain the block diagram in Fig. 5.3a.

The block diagram in Fig. 5.3b is equivalent to that in Fig. 5.3a because addition is a linear operator. In this work, we consider two possible expressions for $h_{ns}[n]$ and $h_{dtr}[n]$. Hence, the block diagram in Fig. 5.3b gives rise to four possible cases; these are summarized in Table 5.1.

In order to determine the maximum difference between $in_{avg}[n]$ and $out_{avg}[n]$ for the cases in Fig. 5.1, we recall that $dtr[n]$ is included in the interval $[-\Delta d/2,\ \Delta d/2]$. Moreover, the quantization errors of a quantizer and a sigma-delta modulator that are defined by Eqs. (5.5) and (5.7) are included in the interval [0, 1]. The outputs

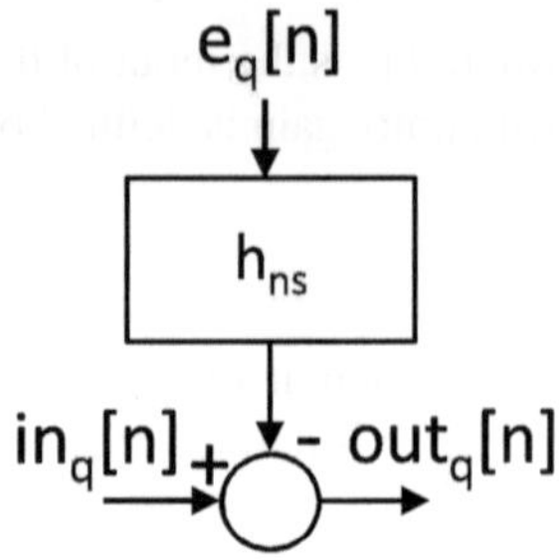

Fig. 5.2 Block diagram that represents the governing equations of a quantizer and a sigma-delta modulator when $h_{ns}[n]$ is given by Eqs. (5.11) and (5.12), respectively

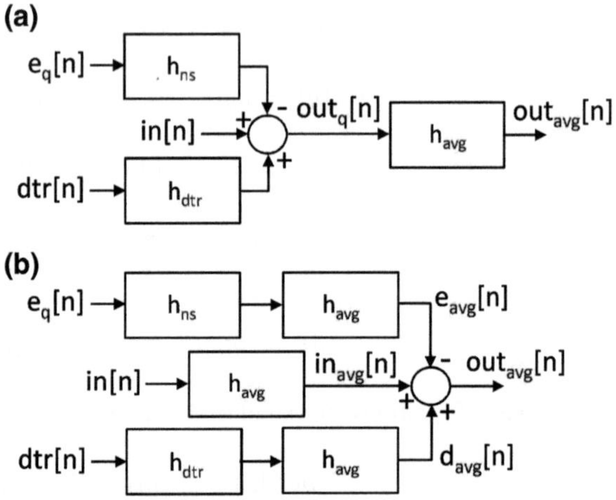

Fig. 5.3 Block diagrams that give the models of **a** a "quantizer plus moving average filter" and **b** a "first-order sigma-delta modulator plus moving average filter" when $h_{ns}[n]$ is given by Eqs. (5.11) and (5.12), respectively

Table 5.1 Possible cases that are obtained from Fig. 5.3b for different expressions of $h_{ns}[n]$ and $h_{dtr}[n]$

$h_{dtr}[n]$	$h_{ns}[n]$	
	$\delta_q[n]$	$\delta_q[n-1] - \delta_q[n]$
$\delta[n]$	Quantizer with zeroth-order shaped dither	First-order sigma-delta with zeroth-order shaped dither
$\delta[n] - \delta[n-1]$	Quantizer with first-order shaped dither	First-order sigma-delta with first-order shaped dither

of the systems $h_{ns}[n]$ and $h_{dtr}[n]$ are equal to their respective inputs in the cases in Table 5.1 that are associated with zeroth-order shaped dither and a quantizer. Hence, for zeroth-order shaped dither and a quantizer, $d_{avg}[n]$ and $e_{avg}[n]$ are equal to the outputs of moving average filters that are fed with $dtr[n]$ and $e_q[n]$. By recalling that the output of a moving average filter is included in the same interval as its input, we conclude that $d_{avg}[n]$ and $e_{avg}[n]$ are included in $[-\Delta d/2, \Delta d/2]$ and [0, 1] for zeroth-order shaped dither and a quantizer, respectively.

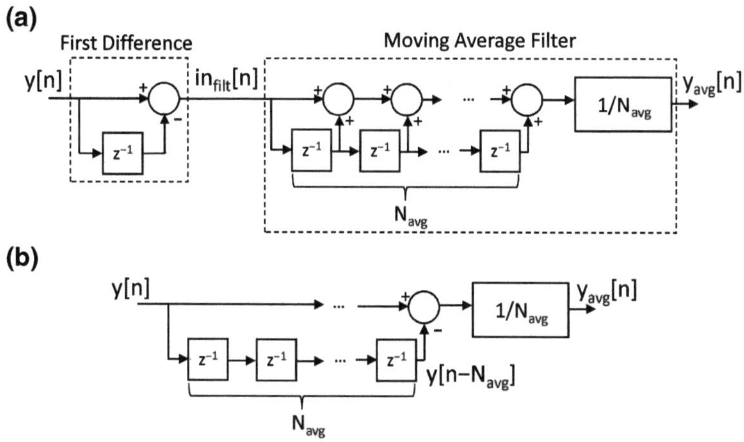

Fig. 5.4 Block diagrams of **a** the system "first difference plus moving average filter" and **b** an equivalent representation with "subtraction plus delay line"

The outputs of the systems $h_{ns}[n]$ and $h_{dtr}[n]$ are equal to the first differences of their respective inputs in the cases in Table 5.1 that are associated with first-order shaped dither and a sigma-delta modulator. Next, we demonstrate that the output of the system "first difference plus moving average filter" lies in the interval $[(y_{min} - y_{max})/N_{avg}, (y_{max} - y_{min})/N_{avg}]$ when the input lies in $[y_{min}, y_{max}]$. We will first prove that the block diagram in Fig. 5.4b is equivalent to that in Fig. 5.4a.

In order to derive the block diagram in Fig. 5.4b from that in Fig. 5.4a, we recall that the output $y_{avg}[n]$ of the moving-average filter in Fig. 5.4a is defined by:

$$y_{avg}[n] = \frac{1}{N_{avg}} \sum_{i=n-N_{avg}+1}^{n} in_{filt}[i], \tag{5.13}$$

where in_{filt} is the input of the moving-average filter and N_{avg} is the number of samples that are used by the filter. From Eq. (5.13) and the block diagram in Fig. 5.4a we can write that the output $y_{avg}[n]$ of the system in Fig. 5.4a is given by:

$$y_{avg}[n] = \frac{1}{N_{avg}} \sum_{i=n-N_{avg}+1}^{n} (y[i] - y[i-1]). \tag{5.14}$$

Since addition is a linear operator, Eq. (5.14) can be rewritten as:

$$y_{avg}[n] = \frac{1}{N_{avg}} \sum_{i=n-N_{avg}+1}^{n} y[i] - \frac{1}{N_{avg}} \sum_{i=n-N_{avg}}^{n-1} y[i]. \tag{5.15}$$

Table 5.2 Maximum difference between $in_{avg}[n]$ and $out_{avg}[n]$

Dither	Noise shaping	
	Zeroth-order (quantizer)	First-order (sigma-delta modulator)
Zeroth-order shaped dither	$1 + \Delta d$	$\frac{2}{N_{avg}} + \Delta d$
First-order shaped dither	$1 + \frac{2\Delta d}{N_{avg}}$	$\frac{2}{N_{avg}} + \frac{2\Delta d}{N_{avg}}$

Equation (5.15) includes two summations that can be written as:

$$\sum_{i=n-N_{avg}+1}^{n} y[n] = \left(\sum_{i=n-N_{avg}+1}^{n-1} y[i] \right) + y[n] \tag{5.16}$$

and

$$\sum_{i=n-N_{avg}}^{n-1} y[i] = \left(\sum_{i=n-N_{avg}+1}^{n-1} y[i] \right) + y[n - N_{avg}], \tag{5.17}$$

respectively. By substituting Eqs. (5.16) and (5.17) into Eq. (5.15), we have:

$$y_{avg}[n] = \frac{1}{N_{avg}} \left(y[n] - y[n - N_{avg}] \right). \tag{5.18}$$

Equation (5.18) is represented by the block diagram in Fig. 5.4b.

From Fig. 5.4b and Eq. (5.18), we conclude that when $y[n]$ is defined in the interval $[y_{min}, y_{max}]$, the output $y_{avg}[n]$ is limited in the interval $[(y_{min} - y_{max})/N_{avg}, (y_{max} - y_{min})/N_{avg}]$. Notice that the size of the interval in which $y_{avg}[n]$ is included is equal to $2(y_{max} - y_{min})/N_{avg}$ and that it decreases linearly with N_{avg}.

We recall that $d_{avg}[n]$ and $e_{avg}[n]$ are the outputs of first difference blocks followed by moving average filters for first-order shaped dither and a sigma-delta modulator. Hence, $d_{avg}[n]$ and $e_{avg}[n]$ are included in the intervals $[-\Delta d/N_{avg}, \Delta d/N_{avg}]$ and $[-1/N_{avg}, 1/N_{avg}]$, respectively, for first-order shaped dither and a sigma-delta modulator.

As can be deduced from Fig. 5.3, the maximum difference between $in_{avg}[n]$ and $out_{avg}[n]$ is equal to the size of the interval in which the quantity $d_{avg}[n] - e_{avg}[n]$ is included. We recall that the random variable $y_3 = y_1 - y_2$ lies in the interval $[y_{1min} - y_{2max}, y_{1max} - y_{2min}]$ when the random variables y_1 and y_2 are included in the intervals $[y_{1min}, y_{1max}]$, $[y_{2min}, y_{2max}]$. The size of the interval in which y_3 is included is equal to the sum of the sizes of those in which y_1 and y_2 are included. Hence, the size in which $d_{avg}[n] - e_{avg}[n]$ is included is equal to the sum of the sizes of the intervals in which $d_{avg}[n]$ and $e_{avg}[n]$ are included. Table 5.2 summarizes the four possible combinations.

From Table 5.2, we conclude that, when N_{avg} is greater than 2, the maximum difference between $in_{avg}[n]$ and $out_{avg}[n]$ in a "first-order sigma-delta modulator plus moving average filter" is always smaller than that in a "quantizer plus moving average filter" with the same value of N_{avg} and the same dither.

5.3 Comparison Between Zeroth and First-Order Shaped Dither

Table 5.2 requires only that the dither and the quantization error are distributed over the intervals $[-\Delta d/2, \Delta d/2]$ and $[0, 1]$, respectively. However, if we assume that the dither $dtr[n]$ is also *uniformly* distributed over $[-\Delta d/2, \Delta d/2]$, we can make some additional important observations. In fact, when $dtr[n]$ is zeroth-order noise shaped and uniformly distributed over $[-\Delta d/2, \Delta d/2]$, the distribution of $d_{avg}[n]$ is approximately Gaussian for N_{avg} "sufficiently large", according to the Central Limit Theorem [2]. We determine by simulation how large a value of N_{avg} is "sufficiently large" to make the Central Limit Theorem valid in the case of a moving average filter. Figure 5.5 shows the distributions of the output of a moving average filter when the input is uniformly distributed over $[-0.5, 0.5]$ for N_{avg} equal to 2, 5 and 10. The dotted and dashed curves in Fig. 5.5 are Gaussian curves fitted to the histograms of the output of the moving average filter when N_{avg} is equal to 5 and 10, respectively. When N_{avg} is equal to 2, the distribution of the output of the moving is triangular. However, when N_{avg} is equal to 5 or 10, the dotted and dashed Gaussian curves fit approximately the histograms of the output of the moving average filter. Hence, the Central Limit Theorem is valid for a moving average filter with uniformly distributed input when N_{avg} is greater than or equal to 5.

We recall that 99.7 % of the values of a random variable with Gaussian distribution are included in the interval $[\mu - 3\sigma, \mu + 3\sigma]$, where μ and σ are the mean value and the standard deviation of the random variable, respectively. Hence, 99.7 % of the values of $d_{avg}[n]$ are included in the interval $[\mu_{d_{avg}} - 3\sigma_{d_{avg}}, \mu_{d_{avg}} + 3\sigma_{d_{avg}}]$ where $\mu_{d_{avg}}$ and $\sigma_{d_{avg}}$ are the mean value and standard deviation of the variance of $d_{avg}[n]$, respectively, when the dither is zeroth-order shaped.

The output of a moving average filter that is fed with $dtr[n]$ has variance (Fig. 5.5):

$$\begin{aligned}\sigma^2_{d_{avg}} &= \mathrm{var}\left(\frac{1}{N_{avg}} \sum_{i=n-N_{avg}-1}^{n} dtr[i]\right)\\ &= \frac{1}{N^2_{avg}} \mathrm{var}\left(\sum_{i=n-N_{avg}+1}^{n} dtr[i]\right) = \frac{\sigma^2_d}{N_{avg}},\end{aligned} \tag{5.19}$$

where σ_d is the square root of the variance of $dtr[n]$. We recall that the square root of the variance of a random variable with uniform distribution over $[-\Delta d/2, \Delta d/2]$ is given by [3]:

$$\sigma_d = \frac{\Delta d}{\sqrt{12}}. \tag{5.20}$$

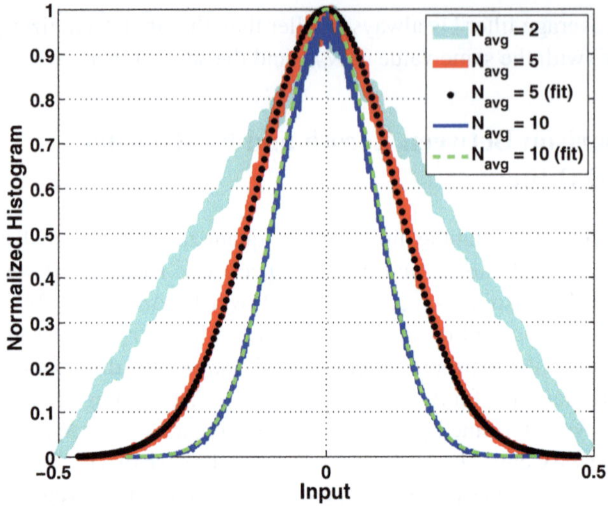

Fig. 5.5 Example distributions of the output of a moving average filter when the input is uniformly distributed over $[-0.5, 0.5]$ for N_{avg} equal to 2, 5 and 10

By substituting Eq. (5.20) into (5.19) we can write:

$$\sigma_{d_{avg}} = \frac{\Delta d}{\sqrt{12 N_{avg}}}. \tag{5.21}$$

As can be seen from Eq. (5.21), the size of the interval in which 99.7 % of the values of $d_{avg}[n]$ are included decreases with $\sqrt{N_{avg}}$ when the dither is zeroth-order noise shaped. Thus, the precision of the measurement given by a quantizer plus moving average filter with zeroth-order shaped dither increases with increasing N_{avg}, even though the maximum difference between $in_{avg}[n]$ and $out_{avg}[n]$ is independent of N_{avg}, as shown in Table 5.2.

It is not necessary to consider the shape of the distribution of $d_{avg}[n]$ when the dither is first-order shaped in order to show the advantages of first-order shaped dither compared to zeroth-order shaped dither. In fact, Table 5.3 compares the intervals in which are included 99.7 and 100 %, respectively, of the values of the dithers after zeroth- and first-order shaping and filtering. For N_{avg} much greater than 10, the size of the interval in which all of the values of d_{avg} are included with first-order shaped dither is always smaller than that in which 99.7 % of the values of d_{avg} are included with zeroth-order shaped dither.

Figure 5.6 shows the sizes of the intervals that are associated with zeroth- and first order noise shaped dither in Table 5.3 for $\Delta d = 1$ and $N_{avg} = 1, 2, \ldots, 30$. As can be seen from Table 5.3, the sizes of the two intervals that are plotted in Fig. 5.6 are scaled by the same quantity Δd. Hence, whether the solid curve is above or below the dotted curve in Fig. 5.6 is independent of Δd.

Table 5.3 Comparison between the intervals in which are included 99.7 % and all the values of the dithers after zeroth- and first-order shaping and filtering, respectively

	Dither after shaping and filtering (d_{avg}) in quantizer and first-order sigma-delta modulator
Zeroth-order shaped dither	99.7 % of the values in $\left[\frac{-3\Delta d}{\sqrt{12N_{avg}}}, \frac{3\Delta d}{\sqrt{12N_{avg}}}\right]$
First-order shaped dither	All of the values in $\left[-\frac{\Delta d}{N_{avg}}, \frac{\Delta d}{N_{avg}}\right]$

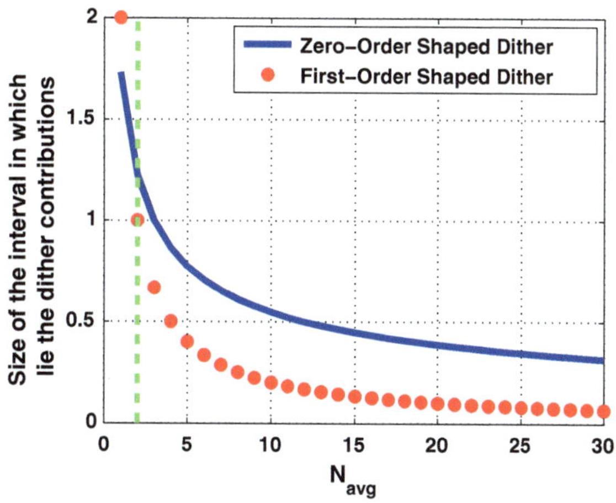

Fig. 5.6 Sizes of the intervals that are associated with zeroth- and first-order noise shaped dither in Table 5.3 for $\Delta d = 1$ and $N_{avg} = 1, 2, \ldots, 30$

The vertical dashed line in Fig. 5.6 is associated with $N_{avg} = 2$. The dotted curve is always below the solid curve for N_{avg} greater than or equal to 2. However, in order to guarantee a Gaussian distribution of the dither contributions with zeroth-order noise shaping, we assume that N_{avg} is greater than or equal to 5. We conclude from Fig. 5.6 that the precisions of a quantizer and a sigma-delta modulator plus moving average filters with first-order shaped dither are higher than those of the same systems with zeroth-order shaped dither when N_{avg} is greater than or equal to 5.

5.3.1 Comparison Between a Quantizer and a First-Order Sigma-Delta Modulator

In this section, we highlight the advantages of first-order noise-shaping of the quantization error with respect to zeroth-order noise shaping. We assume that the quantization error in the system "quantizer plus moving average filter" is uniformly distributed over the interval [0, 1]. By means of the same considerations that we have reported for the distribution of $d_{avg}[n]$ in Sect. 5.3, we conclude that 99.7 % of the values of $e_{avg}[n]$ are included in the interval that is reported in Table 5.4 for a quantizer with zeroth or first-order shaped dither when N_{avg} is larger than 10. Notice that the

Table 5.4 Comparison between the intervals in which are included 99.7 % and all the values of the quantization error after zeroth- and first-order shaping and filtering, respectively

	Quantization error after shaping and filtering (e_{avg})
Quantizer with zeroth and first order shaped dither	99.7 % $\left[0.5 - \frac{3}{\sqrt{12N_{avg}}}, 0.5 + \frac{3}{\sqrt{12N_{avg}}}\right]$
First-order sigma-delta modulator with or without dither	All of the values in $\left[-\frac{1}{N_{avg}}, +\frac{1}{N_{avg}}\right]$

sizes of the intervals that are reported in Table 5.4 are the same of those reported in Table 5.3 when $\Delta d = 1$. Hence, the solid and dotted curves in Fig. 5.6 are equal to the sizes of the intervals in which 99.7 and 100 %, respectively, of the values of e_{avg} are included in a quantizer and in a sigma-delta modulator.

Thus, we conclude that, when N_{avg} is much greater than 10, the size of the interval in which all of the values of e_{avg} are included in a sigma-delta modulator is always smaller than that in which 99.7 % of the values of e_{avg} are included in a quantizer.

We conclude that the precision of a sigma-delta modulator plus moving average filter is higher than that of a quantizer plus moving average filter with the same dither when N_{avg} is larger than 10.

5.4 Conclusion

In this Chapter, we have determined analytically the precisions of the systems "quantizer plus moving average filter" and "first-order noise-shaping sigma delta modulator plus moving average filter" in terms of (i) the maximum difference between the input and the output and (ii) the sizes of the intervals in which are included 99.7 % of the contributions of the dither and the quantization error at the output. We assume that 99.7 % of these contributions represents the most of them. We have shown that the precisions of a quantizer and a sigma-delta modulator plus moving average filters with first-order shaped dither are higher than those of the same systems with zeroth-order shaped dither when N_{avg} is larger than 10. Furthermore, we have demonstrated that the precision of a sigma-delta modulator plus moving average filter is higher than that of a quantizer plus moving average filter with the same dither when N_{avg} is larger than 10.

References

1. S. Haykin, B. Van Veen, "Signals and Systems", Second Edition, *John Wiley & Sons Inc.*, 2003.
2. E. W. Weisstein, "Central Limit Theorem". from *MathWorld*-A Wolfram Web Resource. Available at: http://mathworld.wolfram.com/CentralLimitTheorem.html. Accessed in Nov. 2012.
3. R. Schreier, G. C. Temes, "Understanding Delta-Sigma Data Converters", *IEEE Press, John Wiley & Sons Inc.*, 2005.

Chapter 6
Efficient Modeling and Simulation of Accumulator-Based ADPLLs

6.1 Introduction

In this chapter, we focus on the behavioral modeling and simulation of accumulator-based ADPLLs. First, we introduce some basic concepts related to mixed-signal systems and simulators. We highlight the major issues for the simulation of an ADPLL as an example mixed-signal system. Then we describe how to realize an efficient behavioral model of an ADPLL that produces the minimum number of events during a simulation. We show that, in order to realize an efficient ADPLL model in Simulink, it is necessary to use models that (i) can control the simulation loop during a simulation, and (ii) use predictor equations to assist the simulation. Requirements (i) and (ii) are addressed by means of Matlab S-functions and oscillator models that use equations to predict when some specific edges should occur, respectively.

Afterwards, we describe behavioral models of the remaining building blocks in an ADPLL that are based on Triggered Subsystems and Embedded Matlab Functions.

Finally, we compare the simulation results of an efficient and an improved ADPLL model in terms of the number of samples produced and the total time required to complete the simulation.

6.2 Mixed Signal Systems

Information plays a fundamental role in Electronics. In fact, the purpose of any electronic circuit can be related to processing, storing or transmitting information. In a circuit or an electronic system, information is associated with a signal. The way information is related to a signal is not unique. In the simplest cases, information is associated with the instantaneous value of a current or voltage, but it can also be associated with the duration of a time interval, with the frequency or phase of a periodic signal, or with any feature of a signal that is suitable for the designer of

F. Brandonisio and M. P. Kennedy, *Noise-Shaping All-Digital Phase-Locked Loops*, Analog Circuits and Signal Processing, DOI: 10.1007/978-3-319-03659-5_6,

the circuit. Circuits and systems can be classified as analog, digital, continuous- and discrete-time, depending on how the information and the signals are related to each other. In analog circuits, the values of a signal that can be related to information are continuous. By contrast, in digital circuits only some discrete levels of the signals are considered significant in terms of information. In continuous-time circuits, all the time instants associated with a signal are interesting in terms of information. By contrast, in discrete-time circuits, we consider only a discrete set of time instants. A mixed-signal electronic system is one that includes both analog and digital subsystems.

6.3 Simulation Tools and Solvers of Differential Equations

6.3.1 Continuous-Time and Discrete-Event Simulators

A model is a simplified representation of a real system. In this work, we focus on simulations of models of systems with respect to time, i.e. time-domain simulations. A time-domain simulation simulates a system at a finite number of instants that are distributed over a time interval. When the simulation is running, the current instant at which the system is simulated is often referred to as the "current simulation time" [1, 2]. Moreover, the duration of the time interval between pairs of consecutive instants included in the simulation is called the "time step" [1, 2].

A Simulator uses a "solver" to determine the values of the simulation time and of the signals and the states of the model [3]. Simulators can be separated into continuous-time and discrete-event types [2–4].

In a continuous-time simulator, a set of differential equations describes how the states and outputs of a system evolve in time. The set of differential equations is solved numerically by a continuous-time solver. A continuous-time solver uses a numerical method, that can be referred to as an "integration algorithm", in order to determine the solution of a set of differential equations at each simulation time [3, 5]. By definition, the solution of the differential equations that describe a continuous-time system can comprise samples of the states and signals that are associated with *any* instant. The numerical method determines which instants have to be included in the simulations in order to have a prescribed accuracy.

In a discrete-event simulator, the model of the system describes the relationship between the values of its states and outputs between the current and the next simulation time by means of finite difference equations. By definition, the signals and states of a discrete-time system are defined only over a finite set of discrete-time instants. Moreover, *explicit* finite difference equations do not require a numerical method to be solved. Thus, when the model of the system can be represented by *explicit* finite difference equations, the discrete-time solver of a discrete-event simulator *does not need* a numerical method [3] to update the states and signals of a model or to determine which instants have to be included in the simulation.

For example, we can simulate an oscillator that is applied at the input of an analog filter with a continuous-time simulator. We can use a set of nonlinear differential equations to model the oscillator and use a solver to determine the output of the system "oscillator plus filter".

By contrast, an oscillator that provides the clock to an edge-triggered digital circuit can be simulated with a discrete-event simulator. Changes in the states of the digital circuit occur only at the edges[1] of the clock. Thus, the model of the oscillator could simply generate transitions of its output with a specific period. Only the instants of the edges of the oscillator need to be included in the simulation. Those instants could also be determined directly from the model of the oscillator. Modelling only the transitions of an oscillator is an interesting approach for *behavioral* simulations of mixed systems, as will be shown in the next sections.

6.3.2 Commercial Simulators

Commercial simulators that can be used to simulate and design mixed-signal systems include, for example, Matlab [6], Cadence Design Suite [7] and Xilinx Design Suite [8]. Simulink, which is part of Matlab, can use a continuous- or discrete-time solver, depending on the simulation options. However, it is not possible to use a continuous-time solver and a discrete-time solver at the same time in the same simulation [3, 5].

By contrast, Cadence Virtuoso AMS Designer [7], which is part of Cadence Design Suite, is a single-engine mixed-signal simulator based on the time-continuous Spectre [9] circuit simulator and the discrete-event core of the NC simulator [10]. Spectre and the NC discrete-event core can be used *together* during the simulation of a model that comprises both continuous- *and* discrete-time parts. AMS Designer provides a complete simulation solution for complex mixed-signal circuits [11].

ISim [12] and ModelSim [13], which are included in Xilinx Design Suite, are discrete-event simulators that do not support any mixed-signal features. Simulink and Matlab have their own proprietary programming languages, whereas Cadence AMS Designer supports Verilog-AMS. ISim and ModelSim support only digital HDL languages such as Verilog and VHDL.

In the next sections, we will assume Matlab and Simulink as the reference simulation environments for ADPLLs, which form a particular class of mixed-signal systems.

[1] In the presence of *non-zero* rise and fall times, the "instant" of a transition can be defined as the instant at which the signal crosses a reference (threshold) level.

#	Step	Output
1	System-Level Design	Behavioural models
2	Circuit-Level Design	Models based on industrial libraries
3	Physical Implementation	Layout/Partitioning
4	Physical Verification	LayoutVsSchematic
5	Final Verification	Parasitic extraction

Fig. 6.1 Design flow of a mixed-signal system on chip

6.4 Design of Mixed-Signal Systems on Chip

Figure 6.1 shows the main steps in the design flow of a mixed-signal system on chip [14]. Before the design starts, the designer is committed to some specifications that the system has to meet. The first step of the design flow is System-Level Design. During this step, the designer investigates which system architecture can fulfil the required specifications. For *large mixed-signal* circuits, it is valuable to study the architecture of the system by means of high-level behavioral models for two key reasons. Firstly, the simulation of a behavioral model of the system requires less time than that of a transistor level model. Secondly, a behavioral model can also be modified quickly compared to one at the transistor level.

Furthermore, it is possible to use different approaches when modelling analog and digital parts of a system at the behavioral level. For example, the analog and digital parts can be simulated with differential and difference equations, respectively.

At the end of the system-level design step, a target architecture has been chosen and studied by means of behavioral models. Example tools that can be used to perform the simulations at system-level are Matlab-Simulink, Verilog-AMS, SystemC [15] and CppSim [16].

During the Circuit-Level Design step, a circuit is associated with the target architecture of the system. This process is typically referred to as "synthesis" in tools such as the Cadence or Xilinx Design Kits. The Xilinx ISE can *automatically* synthesize a circuit from the behavioral model of a digital circuit implemented in Verilog. Moreover, the Xilinx System Development Kit (SDK) provides Simulink libraries for the automatic synthesis of Simulink models [17]. Unfortunately, commercial EDA tools cannot yet synthesize general mixed-signal circuits modelled with Verilog-AMS or Matlab code [14].

The output of the Physical Implementation phase is a layout of the circuit. The physical implementation of a mixed-signal circuit is carried out by partitioning the system into analog and digital sections, and realizing the layouts of the analog and digital parts separately.

The Physical and Final Verifications involve running Layout versus Schematic (LVS) simulations and parasitics extraction, respectively, by means of the same tools that are used for the physical implementation.

The specifications are usually met by optimizing the design at a particular level. For example, the design of a small circuit can be optimized at the physical level. However, the designer might realize that a new architecture or new circuit parameters are necessary to achieve the desired degree of optimization at the physical level. Hence, the designer might have to iterate some or all of the steps of the design flow several times, when focusing on the physical level. Iterating the design phases several times is practical only for small circuits. By contrast, it is convenient to optimize the design at the architecture level when dealing with large mixed-signal systems. In some cases, such as transistor level simulation of a PLL, the time required to complete a simulation can be impractically long. In such cases, the design has to be optimized at the architecture level.

In this work, we focus on the architecture level of the design flow of an ADPLL. In particular, we show how to implement efficient behavioral models of ADPLLs that can help the designer to optimize the design at the architectural level.

6.5 Simulation Issues with All-Digital PLLs

All-Digital PLLs are examples of mixed-signal circuits. In fact, the operation of an ADPLL is based on controlling the phase error between two continuous-time signals, namely the output signals of the reference oscillator and the DCO. The phase error is an *analog* quantity because it is proportional to the duration of the continuous (analog) time interval between pairs of consecutive edges of the reference oscillator and the DCO. The main simulation issues for a mixed-signal system that affect an ADPLL are related to the generation of events at fixed instants, to the time precision, and to the quantization of time in the simulator.

6.5.1 Generating Events at Fixed Instants

A time-discrete behavioral model of a digital circuit changes the value of its output at some fixed instants that, in some cases, can be predicted *before* the simulation is executed. For example, consider a general digital circuit that triggers on the rising edges of an oscillator. The instants at which the output of the circuit changes, namely the instants of the rising edges of the oscillator, are known and fixed when the initial phase of the oscillator and the values of all the periods of the oscillator are known and fixed.

By contrast, a solver determines the instants at which the output of the model of an *analog* circuit has to be changed *during* the simulation. When simulating mixed-signal circuits, the solver has to include in the simulations the fixed instants associated with the digital parts of the circuit. How to implement models that can force the solver to generate events at particular instants depends on the simulation platform and on the designer's preferences. A general approach is to implement models that can read the "current simulation time", as we will show in the next sections.

6.5.2 Variation of the Time Step

An analog solver typically changes the size of the time step according to the derivative of the signals in the simulation. When the derivative of a signal increases, the size of the time step is reduced in order to generate a precise solution of the differential equations associated with the circuit. In a mixed-signal simulation, the presence of the model of a digital circuit can affect the size of the time step. Since the output signal of the model of the digital circuit changes at fixed instants, the solver is forced to introduce fixed instants in the solution of the differential equations. Moreover, when the slope of the output signal of the model of a digital circuit is high, the size of the time step may need to be reduced significantly. As a consequence, the number of instants included in the simulation can increase excessively; thus, the time required to complete the simulation can become excessively large.

6.5.3 Quantization of Time

The number of instants that are included in the simulation has to be finite in order to have the simulation completed in a finite time. The instants included in the simulation are selected by the solver when the time step is variable. When the time step is fixed, the values of the instants included in the simulation are fixed. During a simulation with a fixed time step, a model can generate events only at the instants defined by the time step. As a consequence, time is quantized during a simulation with a fixed time step. The quantization error associated with the quantization of time in a simulation has to be considered in a model that produces events at specific instants. For example, consider a model that produces a number of consecutive events when the size of the time step is small enough. The same model might wrongly produce only one event when the size of the time step is too large.

ADPLLs are affected by peculiar simulation issues related to the long time required to complete the simulation, and to the presence of multiple clocks.

6.5.4 Long Time Required to Complete a Simulation

The long time that the simulation of a PLL can require is one of the main simulation issues of PLLs. This problem is related to the fact that the bandwidth of a PLL is typically much smaller than the reference frequency, which can be several orders of magnitude smaller than the frequency of the DCO. Thus, in order to run a transient

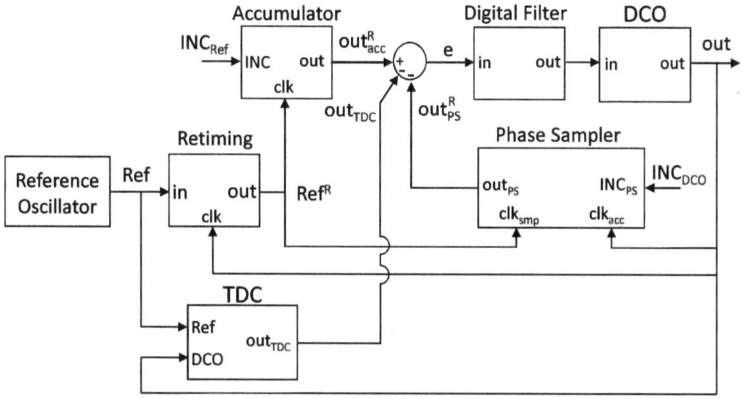

Fig. 6.2 Architecture of an accumulator-based ADPLL with retiming and TDC

simulation in which the PLL model reaches steady state, it can be necessary to simulate hundreds of cycles of the reference oscillator, and hundreds of thousands of cycles of the DCO. For example, the simulation of a PLL at architecture and transistor levels could take days and months, respectively.

Furthermore, there might be issues associated with the numerical overflow of the registers that are used to store the values of the phases, which accumulate in time, of the signals in the model of a PLL.

Hence, in order to study PLLs by means of simulation with acceptable completion times, it is important to develop efficient behavioral models, as we will discuss in Sect. 6.6.

6.5.5 Multiple Clocks and Metastability

At least two oscillators, namely the reference oscillator and the DCO, coexist in an ADPLL. The operations of some parts of the ADPLL, such as the phase-sampler and the TDC, are determined by the edges of these two clocks. When the edges of the two clocks coincide, a metastability issue occurs. In terms of simulation, the metastability problem can be solved by implementing models that detect when a pair of edges is overlapped and *always* give priority to the edge of the same clock when this situation arises.

6.6 Efficient Model of an Accumulator-Based ADPLL

Figure 6.2 shows the architecture of an accumulator-based ADPLL. The phase error e and the output of the digital filter are known once the outputs Ref^R and out_{ps}^R of the accumulator and phase sampler have been determined. Let us define the *significant* edges of the DCO as the first rising edges of the DCO that occur after each of those of

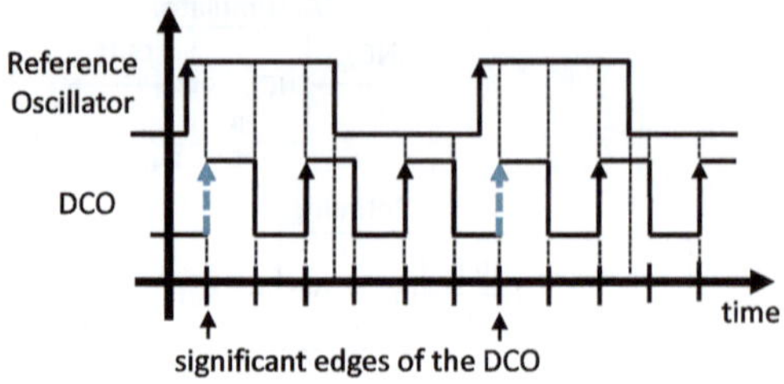

Fig. 6.3 Example timing diagram with the significant edges of a DCO

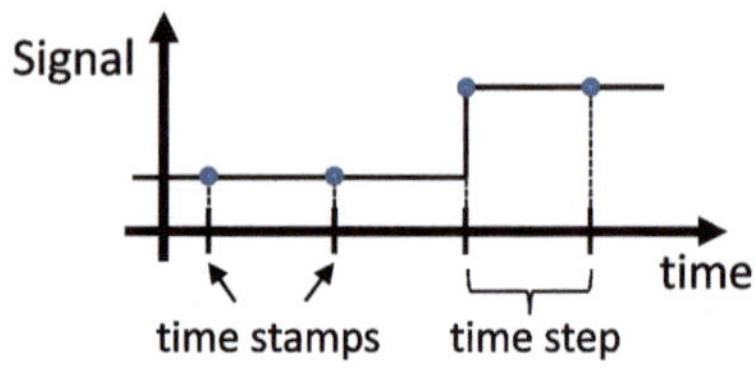

Fig. 6.4 Example time stamps and time steps associated with a signal during a simulation

the reference oscillator. The outputs of the accumulator and phase sampler in Fig. 6.2 are known once the edges of the reference oscillator and the significant edges of the DCO have been determined. Thus, *all* signals in the behavioral model of an ideal ADPLL are known when the instants of the rising edges of the reference oscillator and the significant edges of the DCO have been determined (Fig. 6.3).

Assume that events are associated with the rising edges of the reference oscillator and the DCO. Consider a time-domain simulation of an ADPLL that spans the time interval $[0, t]$. The *minimum* number of events associated with the behavioral simulation of an ADPLL over the time interval $[0, t]$ is given by the number of edges of the reference oscillator and the significant edges of the DCO that are included in the time interval $[0, t]$. The *fastest* possible way to simulate an ADPLL at the behavioral level is to use a model that produces the minimum number of events over the time interval $[0, t]$. The number of events that are generated during a transient simulation is typically determined by the solver of differential equations that is associated with the simulation. The solver also sets the accuracy with which the events are localized in time. In fact, a transient simulation of the model of a system produces samples of the signals and states of the system over the time interval $[0, t]$. A transient simulation with a finite memory has to produce a finite number of samples of the signals and states of a model. A sample is produced by an "event" that occurs during the simulation. An event usually occurs when the value of a signal or a state of the simulated system changes. However, events can also be produced when the signals and states of the simulated system are constant. Let us define a "time stamp" as the instant associated with an event. The difference between two consecutive time stamps is typically called a "time step" [2], as shown in Fig. 6.4.

When the time step is fixed, time is quantized with a resolution equal to the duration of the time step. Time stamps are integer multiples of the time step in a fixed-step simulation. Consider an event that is scheduled to occur at the instant t in a fixed-step simulation. The simulator reports an event at the first time stamp after

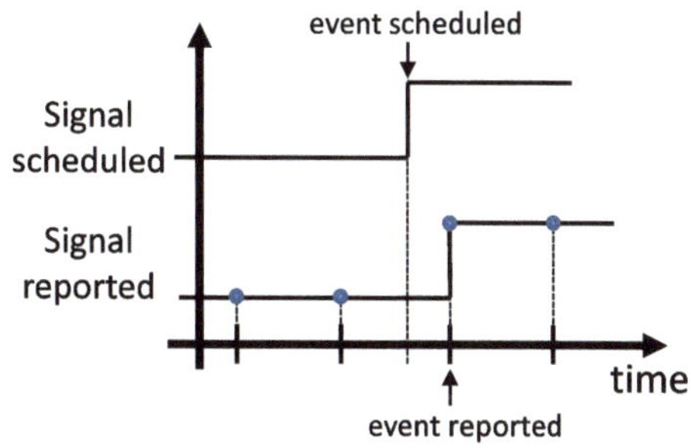

Fig. 6.5 Scheduled and reported signals in a fixed-step simulation

t when t is not equal to a time stamp [2]. Thus, time stamps are produced with a precision equal to the size of the time step in a fixed-step simulation. The number of events of the simulation of a system over the time interval $[0, t]$ with fixed-step t_{step} is given by ceil (t/t_{step}) (Fig. 6.5).

Time stamps can be produced with *machine precision* in a variable time-step simulation. The number of events in a variable time-step simulation depends on the solver. The solver typically reduces the time step in the presence of discontinuities of the signals or states of a system [2]. When a model produces a signal with discontinuities, such as a square wave, the simulation with a variable-step solver can produce a larger number of events compared to a fixed time-step simulation.

In Simulink, the user can schedule events, as explained hereafter. The simulation loop of a Simulink model is illustrated in Fig. 6.7. During the simulation loop, the functions `mdlProcessParameters`(·), `mdlGetTimeOf-NextVarHit`(·), `mdlOutputs`(·), `mdlUpdate`(·) are executed. We will refer to these functions as *mdl-functions*. The mdl-functions are defined in the mex-file that is produced by an S-function builder.

An S-function is a Simulink block that is associated with custom C-code[2] that is written by the user. The simplest way to create an S-function is by means of the Simulink block "S-function builder". An S-function builder provides a graphical interface for describing a model by means of C-code. An S-function builder produces a "Matlab Executable File" (mex-file) from the C-code written by the user. The model that is associated with the mex-file can be used in a Simulink simulation by means of an S-function block in which the name of the mex-file is specified. The mex-files that are associated with the S-functions of a Simulink model have to be included in the current directory. The standard procedure to produce a mex-file is summarized in Fig. 6.6.

The mdl-functions determine the model of the Simulink block that is associated with a mex-file.

The function `mdlProcessParameters`(·) is called if a parameter of the model changes during the simulation. `mdlGetTimeOfNextVarHit`(·) is executed when

[2] In the "Configuration Parameters" menu of a Simulink model, it is possible to select ANSI C or C++. In this work, we selected the ANSI C language with a Generic-Real-Time target hardware.

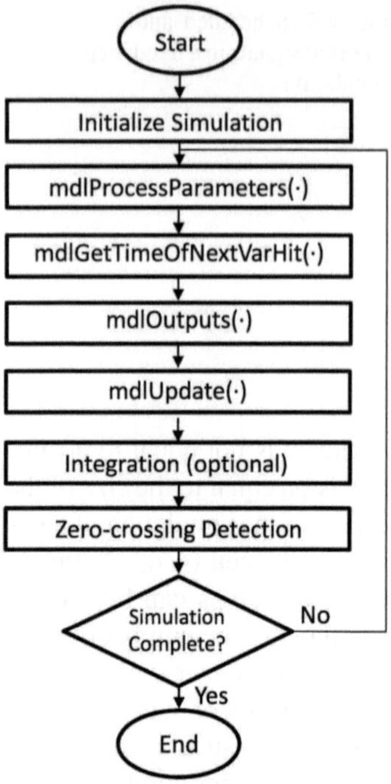

Fig. 6.6 Flow chart of the simulation loop of a Simulink model

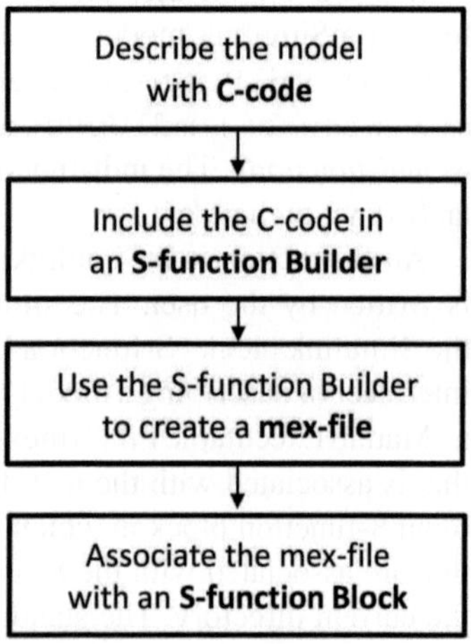

Fig. 6.7 Standard procedure to create the mex-file on an S-function from C-code

the time step of the simulation is variable. `mdlOutputs(·)` and `mdlUpdate(·)` set the outputs and the states of the model, respectively.

A Simulink simulation uses an integration algorithm (labelled "Integration" in Fig. 6.6) when the model includes *continuous states*. The integration algorithm is not executed when the simulation includes models with only discrete states. Finally, a

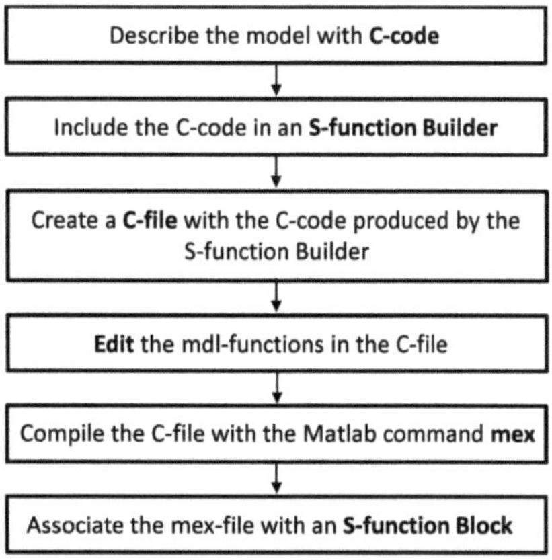

Fig. 6.8 Procedure to create the mex-file on an S-function with custom mdl-functions

Simulink zero-crossing detection algorithm might be used to locate the discontinuities and end the current time step when a zero crossing is detected.

The C-code that defines the mdl-functions of an S-function is accessible from the graphical interface of an S-function builder. An S-function builder *doesn't* allow the user to create a C-file that is based on *modified custom* versions of these functions. However, the user can create a C-file that includes modified mdl-functions and can produce a mex-file by means of the Matlab command *mex*. Figure 6.8 shows the procedure to create mex-files based on custom mdl-functions.

By modifying the mdl-functions, it is possible to achieve precise control of Simulink's simulation loop. For example, the S-function associated with the function `mdlGetTimeOfNextVarHit`$(\cdot)$ in List. 6.1 is executed only at the simulation instant $t = 1$.

```
#define MDL_GET_TIME_OF_NEXT_VAR_HIT
static void mdlGetTimeOfNextVarHit(SimStruct  *S)
{real_T t = 1;
ssSetTNext(S,t);}
```

Listing 6.1 Example mdl-function that schedules the execution of a block for the instant $t = 1$ in Simulink

Even though we consider that the operation of an ADPLL is determined by the rising edges of the reference oscillator and the DCO, we will show how to implement efficient models of the building blocks of an ADPLL that produce events on the rising *and* falling edges of the reference oscillator and the DCO in order to make our examples more general (Fig. 6.8).

6.7 Efficient Model of a Reference Oscillator

Figure 6.9a shows the samples of an example output signal that is produced by the simulation of an example oscillator when the time step is fixed. By contrast, Fig. 6.9b illustrates the samples of the output signal that the simulation of an example *efficient*

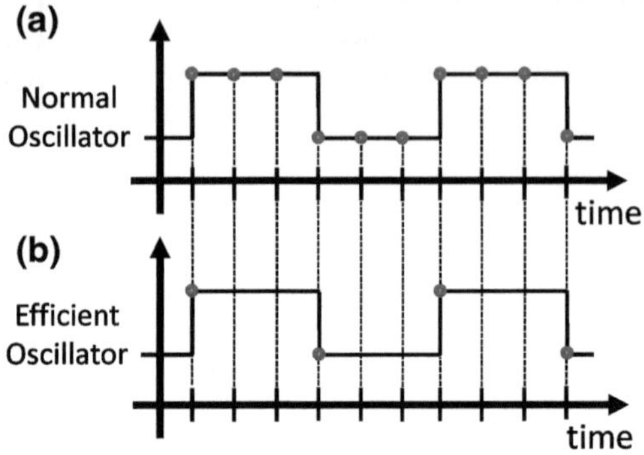

Fig. 6.9 Example samples associated with the simulated output signals of **a** a normal model and **b** an efficient model of an oscillator

model produces. The samples in Fig. 6.9b are associated only with the *transitions* of the oscillator output.

In order to realize the efficient model of an oscillator, it is necessary to produce events at specific instants during a simulation. Events can be controlled in Simulink by means of an S-function, as shown in Sect. 6.6. In fact, the user can specify, by means of the function `ssSetNext`(·), at which simulation instants an S-function is executed. Simulink produces an event every time an S-function is executed. Thus, when the execution of an S-function is scheduled for the instants $t[1]$, $t[2]$,..., $t[n]$, the simulation includes events at the instants $t[1]$, $t[2]$,..., $t[n]$. A simulation that includes *only* an S-function that uses `ssSetNext`(·) to schedule its own execution at the instants $t[1]$, $t[2]$,..., $t[n]$, includes *only* events at the instants $t[1]$, $t[2]$,..., $t[n]$. As a consequence, the efficient model of an oscillator can be implemented by means of an S-function that schedules its own execution for the instants that are associated with the rising and falling edges of the oscillator clock. The simplest representation of a rising or falling edge is the transition of a signal between two levels. Hence, in order to produce a rising or falling edge every time the model is executed, the output of the oscillator can be swapped between two reference values, such as 0 and 1, for example.

We assume an oscillator that produces a signal with a duty cycle equal to 50 %. In this case, the time step of the simulation of an efficient oscillator model is determined by the period of the oscillator divided by two. The time step is fixed or variable when the period of the oscillator is fixed or variable, respectively. The flow chart in Fig. 6.10 summarizes how the simulation time is determined during the simulation of an efficient oscillator model.

In Fig. 6.10, $t[n]$, T and $t_e[n]$ are the simulation time, the period and the instants of the rising and falling edges of the oscillator, respectively.

Figure 6.11 shows the block diagram of an example efficient model of an oscillator based on an S-function. A "Constant" Simulink block that is labelled "Frequency" in Fig. 6.11 sets the value of the frequency of the oscillator. The frequency of the oscillator is set by the input in order not to force the user to compile a new model every time he or she needs to change the frequency of the oscillator. The output

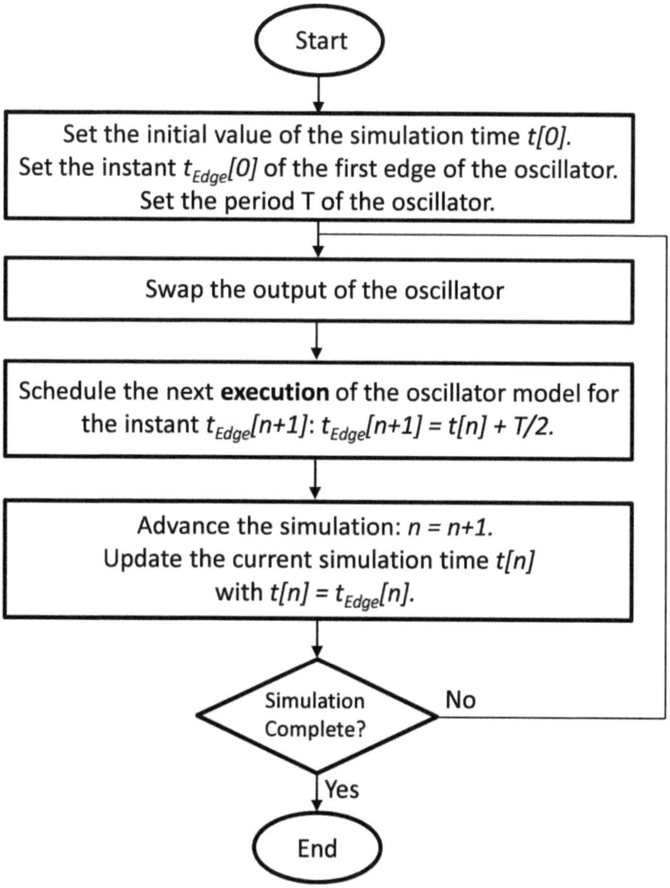

Fig. 6.10 Flow chart of the simulation of an efficient oscillator model

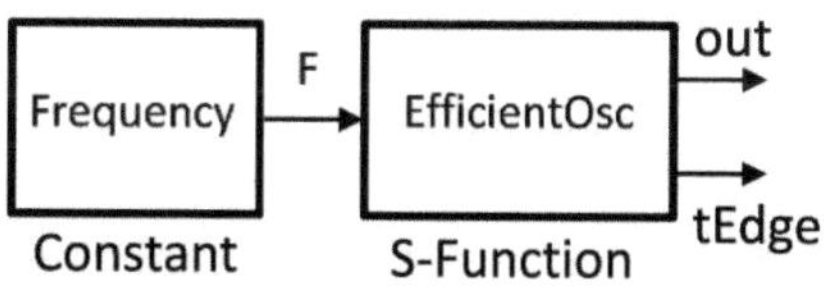

Fig. 6.11 Block diagram of an example efficient oscillator model implemented in Simulink

`tEdge` is equal to the simulation time associated with every edge of the oscillator. `tEdge` is useful for simplifying the behavioral models of other building blocks on an ADPLL, such as the TDC and DCO.

Listings 6.2–6.4 show example mdl-functions that can be used to generate the S-function in Fig. 6.11 by following the procedure in Sect. 6.6.

```
#define MDL_GET_TIME_OF_NEXT_VAR_HIT
static void mdlGetTimeOfNextVarHit(SimStruct *S)
{const real_T *F = (const real_T*)
  ssGetInputPortSignal(S,0);
real_T T; // Period of the oscillator
T = 1/F[0];
ssSetTNext(S, ssGetT(S)+T/2);}
```

Listing 6.2 mdl-function that schedules the execution of the efficient model of an oscillator in Simulink

Listing 6.2 illustrates the function `mdlGetTimeOfNextVarHit` that schedules the execution of the oscillator model by means of the function `ssSetTNext`. The C-code in List. 6.3 updates the outputs *out* and *tEdge* of the oscillator model in Fig. 6.11. Every time the oscillator model is executed, *out* and *tEdge* assume the values of the discrete states `xD[0]` and `xD[1]` of the model.

```
static void mdlOutputs(SimStruct *S, int_T tid)
{real_T *out  = (real_T *)
  ssGetOutputPortRealSignal(S,0);
real_T *tEdge  = (real_T *)
  ssGetOutputPortRealSignal(S,1);
const real_T *xD = ssGetDiscStates(S);
out[0] = xD[0];
tEdge[0] = xD[1];}
```

Listing 6.3 mdl-function that updates the output of the efficient model of an oscillator in Simulink

The states `xD[0]` and `xD[1]` are set by the C-code in List. 6.4. The state `xD[0]` toggles its value (from 0 to 1 and from 1 to 0) every time the model of the oscillator is executed. The state `xD[1]` is equal to the instant at which the execution of the model of the oscillator is scheduled.

```
#define MDL_UPDATE
static void mdlUpdate(SimStruct *S, int_T tid)
{real_T *xD = ssGetDiscStates(S);
const real_T *F = (const real_T*)
   ssGetInputPortSignal(S,0);
real_T T; // Period of the oscillator
T = 1/(F[0]);
if (xD[0] == 0)
{xD[0] = 1;}
else{xD[0] = 0;}
xD[1] = ssGetT(S)+T/2;}
```

Listing 6.4 mdl-function that updates the states of the efficient model of an oscillator in Simulink

6.8 Simple Efficient DCO Model

The modeling approach that has been described in Sect. 6.7 can also be used to create a *simple* efficient model of a DCO. In fact, a simple efficient DCO model can be obtained by modifying the oscillator model in Sect. 6.7 in order to have the period

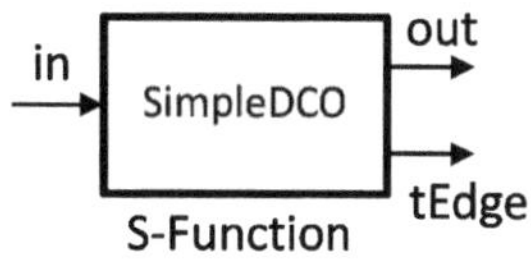

Fig. 6.12 Block diagram of an example S-function that is associated with a simple efficient model of a DCO

T of the DCO given by:

$$T = \frac{1}{f_0 + K_v \cdot in}, \tag{6.1}$$

where f_0, K_v and in are the free-running frequency, gain and input of the DCO, respectively.

The block diagram of a simple efficient DCO model that is based on a Simulink S-function is illustrated in Fig. 6.12.

The input *in* in Fig. 6.12 represents the input digital word of the DCO. The output $tEdge$ represents the instants associated with the DCO edges. $tEdge$ will be used to simplify the behavioral models of a TDC, as will be shown in the next sections.

The mdl functions in Lists. 6.3, 6.5 and 6.6 can be used to generate the S-function in Fig. 6.12, as explained in Sect. 6.6.

```
#define MDL_GET_TIME_OF_NEXT_VAR_HIT
static void mdlGetTimeOfNextVarHit(SimStruct *S){
    const real_T *in = (const real_T*)
   ssGetInputPortSignal(S,0);
real_T T; // Period of the DCO
real_T F0 = 1; // Free-running Frequency
real_T Kv = 10;// Gain
T = 1/(Kv * in[0]+ F0);
ssSetTNext(S, ssGetT(S)+T/2);}
```

Listing 6.5 mdl-function that schedules the execution of the simple efficient model of a DCO in Simulink

```
#define MDL_UPDATE
static void mdlUpdate(SimStruct *S, int_T tid){
real_T *xD  = ssGetDiscStates(S);
const real_T *in = (const real_T*)
  ssGetInputPortSignal(S,0);
real_T T; // Period of the oscillator
real_T F0 = 1; // Free-running Frequency
real_T Kv = 10; // Gain
T = 1/(Kv * in[0]+ F0);
if (xD[0] == 0)
{xD[0] = 1;}
else{xD[0] = 0;}
xD[1] = ssGetT(S)+T/2;}
```

Listing 6.6 mdl-function that updates the states of a simple efficient DCO model in Simulink

Fig. 6.13 Example events that are produced by simulating an efficient reference oscillator, a simple efficient DCO and an improved efficient DCO

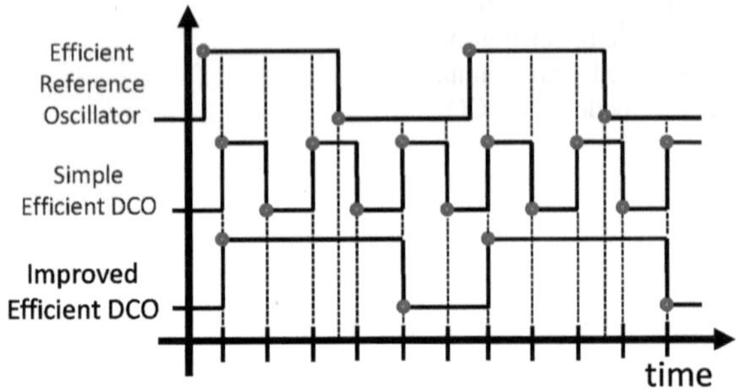

It is important to note that the simulation of an ADPLL that produces the *minimum* number of events includes only the *significant* edges of the DCO, as explained in Sect. 6.6. A simple efficient DCO model produces transitions of the output for *all* edges of the DCO. Thus, the simulation of an ADPLL that includes a simple efficient DCO model doesn't produce the minimum number of events. In the next section, we show how to realize an efficient DCO model that also minimizes the number of events in the simulation of an ADPLL.

6.9 Improved Efficient DCO Model

An improved efficient DCO model that minimizes the number of computed events in the simulation of an ADPLL produces only rising edges that are coincident with the significant edges of a DCO.

Figure 6.13 shows example events that are produced by simulating an efficient reference oscillator, a simple efficient DCO, and an improved efficient DCO. An improved efficient DCO model schedules the output transitions by *calculating* the instants that are associated with the significant edges of the DCO. The instant $t_{eDCO}[n]$ of the nth significant edge of the DCO can be calculated by means of the following recursive equation:

$$t_{eDCO}[n+1] = t_{eDCO}[n] + T_{DCO}[n] \cdot R_{tot}[n], \tag{6.2}$$

where n is the discrete time domain index of the edges of the reference oscillator, $T_{DCO}[n]$ is the period of the DCO which we assume is constant between two consecutive edges of the reference oscillator, and $R_{tot}[n]$ is the number of complete DCO periods with duration $T_{DCO}[n]$ in the interval $[t_{eDCO}[n], t_{eDCO}[n+1]]$. $R_{tot}[n]$ is given by:

$$\tilde{R}_{tot}[n] = \tilde{R}[n] + 1. \tag{6.3}$$

In Eq. (6.3), $\tilde{R}[n]$ is defined by:

$$\tilde{R}[n] = \text{floor}\left(\frac{T_{ref} - T_{DCO}[n-1] + T_{eq}[n-1]}{T_{DCO}[n]}\right), \tag{6.4}$$

where $T_{eq}[n]$ is:

$$T_{eq}[n] = T_{DCO}[n] \cdot \left(R[n] - \tilde{R}[n]\right). \tag{6.5}$$

$R[n]$ in Eq. (6.5) is defined as follows:

$$R[n] = \frac{T_{ref} - T_{DCO}[n-1] + T_{eq}[n-1]}{T_{DCO}[n]}. \tag{6.6}$$

Now, let us demonstrate Eqs. (6.2)–(6.6).

Example timing diagrams of the DCO period and of the edges of the reference oscillator and DCO are illustrated in Fig. 6.14.

We assume that the periods T_{Ref} and T_{DCO} of the reference oscillator and DCO are constant and variable, respectively. We also assume that $T_{DCO}[n] < T_{Ref}[n]$, i.e., there is always at least one edge of the DCO between two consecutive edges of the reference oscillator.

The input to the DCO can change on every edge of the reference oscillator, as shown in Fig. 6.14. The time discrete variable n is defined by the instants when the edges of the reference oscillator occur. We assume that $t_{Ref}(n)$ is the instant at which the nth edge of the associated reference oscillator occurs. We assume that the duration of the DCO period that includes the instant $t_{Ref}[n]$ is still equal to the value of the DCO input that was set at $t_{Ref}[n-1]$.[3] After the DCO period that includes the time instant $t_{Ref}[n]$ is complete, the duration of the DCO period is determined by the DCO input at $t_{Ref}[n]$.

From the timing diagrams in Fig. 6.14, we can write:

$$T_{eq}[n-1] + T_{Ref} + T_{DCO}[n] - T_{eq}[n] = \tilde{R}_{tot}[n] \cdot T_{DCO}[n] + T_{DCO}[n-1], \tag{6.7}$$

where $T_{eq}[n]$ is the duration of the time interval that is defined by the edge of the reference oscillator associated with n and by the next significant edge of the DCO. In Eq. (6.7), $\tilde{R}_{tot}[n]$ is the number of complete DCO periods with duration $T_{DCO}[n]$ that are included in the time interval $[t_{eDCO}(n), t_{eDCO}(n+1)]$. $\tilde{R}_{tot}[n]$ is equal to the ratio between the duration of the time interval $[t_{eDCO}(n), t_{eDCO}(n+1)]$ and $T_{DCO}[n]$. As can be seen from the timing diagram in Fig. 6.14, $\tilde{R}_{tot}[n]$ is given by:

$$\tilde{R}_{tot}[n] = \frac{T_{ref} + T_{DCO}[n] - T_{eq}[n] - T_{DCO}[n-1] + T_{eq}[n-1]}{T_{DCO}[n]}. \tag{6.8}$$

[3] This assumption is equivalent to assuming that the step response of the DCO is a step signal and that the input of the DCO can change only on the edges of the DCO.

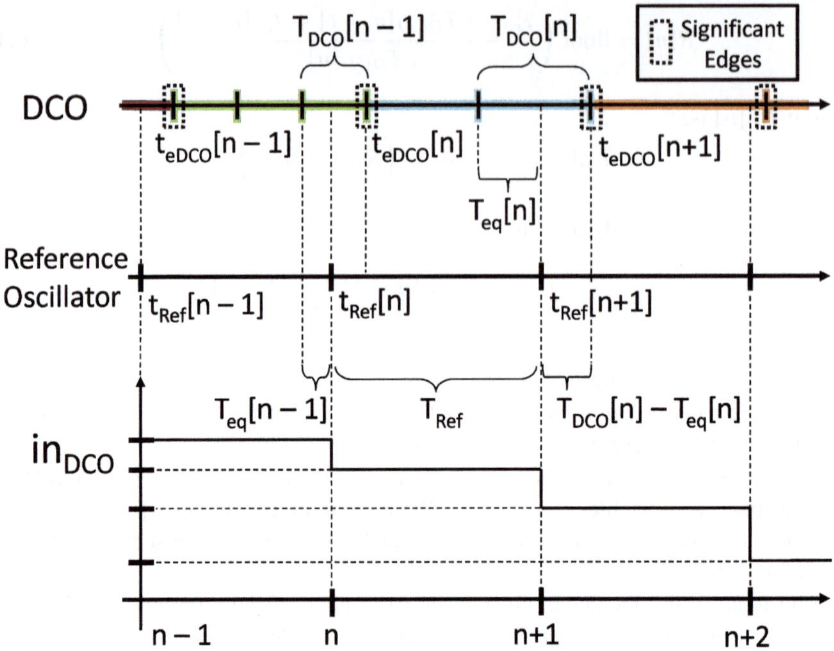

Fig. 6.14 Example period of the DCO and timing diagrams of the rising edges of the reference oscillators and DCO. The periods of the DCO and reference oscillator are variable and constant, respectively

By the definition of a significant DCO edge, the number $R_{tot}[n]$ of complete DCO periods with duration $T_{DCO}[n]$ that are included in the interval $[t_{eDCO}(n), t_{eDCO}(n+1)]$ is equal to that in the interval $[t_{eDCO}(n), t_{Ref}(n+1)]$[4] plus one. The number $\tilde{R}[n]$ of complete DCO periods with duration $T_{DCO}[n]$ that are included in the interval $[t_{eDCO}(n), t_{Ref}(n+1)]$ is equal to the *integer part* of the ratio between the duration of the interval $[t_{eDCO}(n), t_{Ref}(n+1)]$ and $T_{DCO}[n]$. Thus, from Fig. 6.14, $\tilde{R}[n]$ can be defined as follows:

$$\tilde{R}[n] = \text{floor}\left(\frac{T_{ref} - T_{DCO}[n-1] + T_{eq}[n-1]}{T_{DCO}[n]}\right). \tag{6.9}$$

Hence, we can write:

$$\tilde{R}_{tot}[n] = \tilde{R}[n] + 1. \tag{6.10}$$

By substituting Eq. (6.10) into Eq. (6.7), we have:

[4] We assume that the edges of the reference oscillator and the DCO cannot overlap, and that there is always a minimum delay between the edges of the reference oscillator and the significant DCO edges. In particular, we assume that the edges of the reference oscillator occur *before* the significant edges of the DCO when they are very close. This assumption is exact in an ADPLL with retiming.

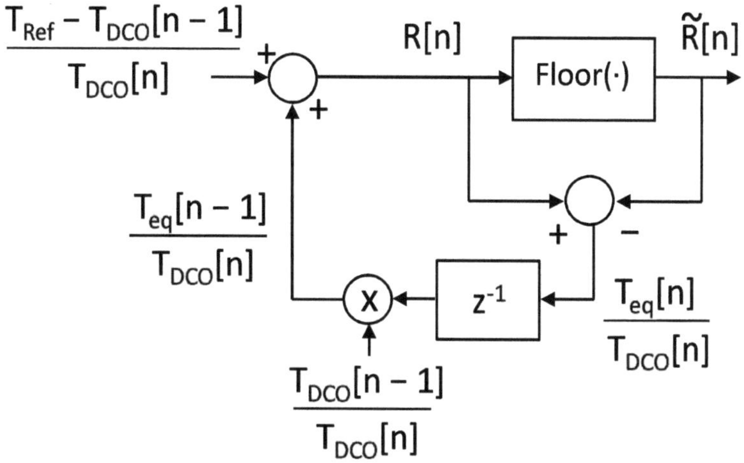

Fig. 6.15 Block diagram representation of the equations to calculate the instants of the significant edges of a DCO

$$\frac{T_{eq}[n]}{T_{DCO}[n]} = R[n] - \tilde{R}[n], \tag{6.11}$$

where $R[n]$ is defined by:

$$R[n] = \frac{T_{ref} - T_{DCO}[n-1] + T_{eq}[n-1]}{T_{DCO}[n]}. \tag{6.12}$$

The block diagram in Fig. 6.15 represents Eq. (6.11). Note that the model in Fig. 6.15 is a *first-order sigma-delta modulator*.

Finally, from Fig. 6.14, we can write:

$$t_{eDCO}[n+1] = t_{eDCO}[n] + T_{DCO}[n]\tilde{R}_{tot}[n]. \tag{6.13}$$

We conclude that it is possible, by means of Eqs. (6.9)–(6.13), to calculate the instant $t_{eDCO}[n+1]$ of the next significant DCO edge from the period T_{Ref} of the reference oscillator, the instant $t_{eDCO}[n]$ of the current significant DCO edge, and the durations $T_{DCO}[n]$ and $T_{DCO}[n-1]$ of the current and previous DCO periods.

An example Simulink implementation of the system "efficient reference oscillator plus improved efficient DCO" is shown in Fig. 6.16. The block labelled "EfficientOsc" in Fig. 6.16 is an efficient model of the reference oscillator that can be implemented with the procedure shown in Sect. 6.7.

Listings 6.7 and 6.8 include example mdl-functions that can be used to create the S-function that is labelled "EfficientDCO" in Fig. 6.16.

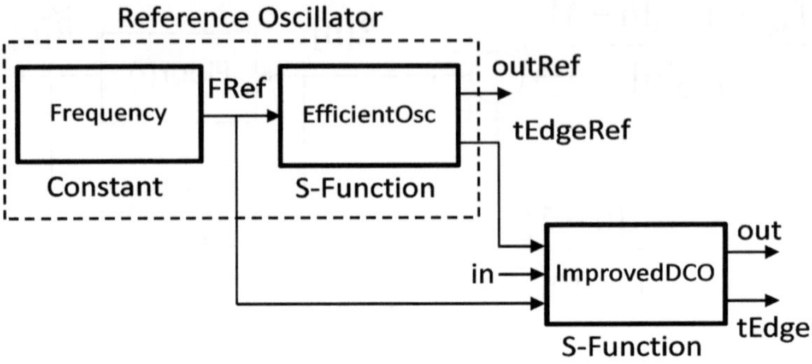

Fig. 6.16 Block diagram of an example system "efficient reference oscillator plus improved efficient DCO" based on S-Functions

```
# define MDL_GET_TIME_OF_NEXT_VAR_HIT
static void mdlGetTimeOfNextVarHit(SimStruct *S)
{const real_T   *in   = (const real_T*)
  ssGetInputPortSignal(S,0);
const real_T   *FRef   = (const real_T*)
  ssGetInputPortSignal(S,1);
real_T    *xD = ssGetDiscStates(S);
real_T TDCO; // Period of the DCO
real_T TRef;// Period of the Reference Oscillator
real_T F0 = 100000000; // Free-running Frequency
real_T Kv = 1000000000; // Gain
real_T Teq;
real_T Rtot ;
real_T R;
real_T Rtilde ;
real_T HalfDCOperiod;
TDCO = 1/( Kv * in [0]+ F0);
TRef = 1/ FRef [0];
R = ( TRef - xD [2] + xD [3]) /TDCO ;
Rtilde = floor(R);
Teq = TDCO * (R - Rtilde);
Rtot = Rtilde + 1;
// xD [1] is t[n +1]; ssGetT (S) is t[n];
// xD [2] is TDCO [n -1]; xD [3] is Teq[n 1];
ssSetTNext (S, ssGetT (S)+ TDCO * Rtot/2 );}
```

Listing 6.7 mdl-function that schedules the execution of an improved efficient DCO model in Simulink

Listings 6.7 and 6.8 implement Eqs. (6.2)–(6.6) in order to schedule the execution of the S-function "EfficientDCO" and to update its states. The state variables `xD[0]` and `xD[1]` are equal to the outputs `out` and `tEdge`, respectively. The state variables `xD[2]` and `xD[3]` are equal to $T_{DCO}[n-1]$ and $T_{eq}[n-1]$, respectively.

```
static void mdlUpdate(SimStruct *S, int_T tid)
{real_T            *xD  = ssGetDiscStates(S);
const real_T   *in  = (const real_T*)
  ssGetInputPortSignal(S,0);
const real_T   *FRef  = (const real_T*)
  ssGetInputPortSignal(S,1);
real_T TDCO; // Period of the DCO
real_T TRef; // Period of the Reference Oscillator
real_T F0 = 100000000;// Free-running Frequency
real_T Kv = 1000000000; // Gain
real_T Teq;
real_T Rtot ;
real_T R;
real_T Rtilde ;
real_T HalfDCOperiod;
TDCO = 1/( Kv * in [0]+ F0);
TRef = 1/ FRef [0];
R = ( TRef - xD [2] + xD [3]) /TDCO ;
Rtilde = floor(R);
Teq = TDCO * (R - Rtilde);
Rtot = Rtilde + 1;
xD [1] = ssGetT (S)+ TDCO * Rtot/2;
if (xD[0] == 0){
xD [2] = TDCO;
xD [3] = Teq;
}
if (xD [0] == 0){xD [0] = 1;}
else {xD [0] = 0;}}
```

Listing 6.8 mdl-function that updates the states of an improved efficient DCO model in Simulink

6.10 Modeling Strategy for the Other Building Blocks of an ADPLL

In Sects. 6.7–6.9, we modelled the reference oscillator and the DCO with S-functions in order to control when events are produced during a Simulink simulation. The other building blocks of an ADPLL that we still have to model are the retiming block, the phase sampler, the accumulator, the digital filter, and the TDC. The edges of the reference oscillator and the DCO determine when the inputs and the states of the other building blocks of the ADPLL change. By implementing behavioral models of the other building blocks on an ADPLL that *do not schedule* events, it is possible to minimize the total number of events that are produced during the simulation of an ADPLL. In order to prevent the models of the other building blocks of an ADPLL from scheduling events, we can use Simulink triggered subsystems. Figure 6.17a and b show a Simulink triggered subsystem that is driven by a signal labelled “Clock”, and an Embedded Matlab Function, respectively.

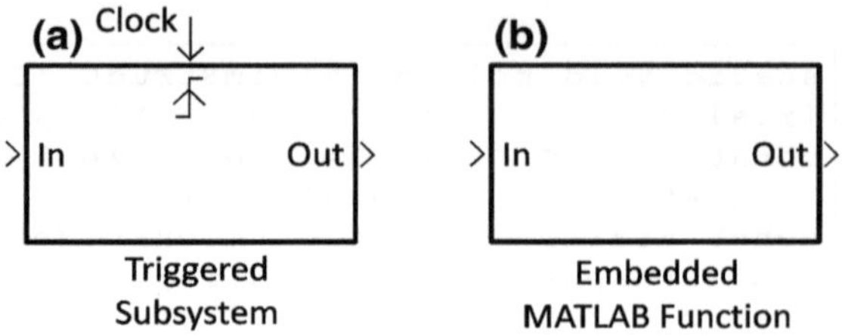

Fig. 6.17 **a** A triggered Subsystem and **b** an Embedded Matlab Function

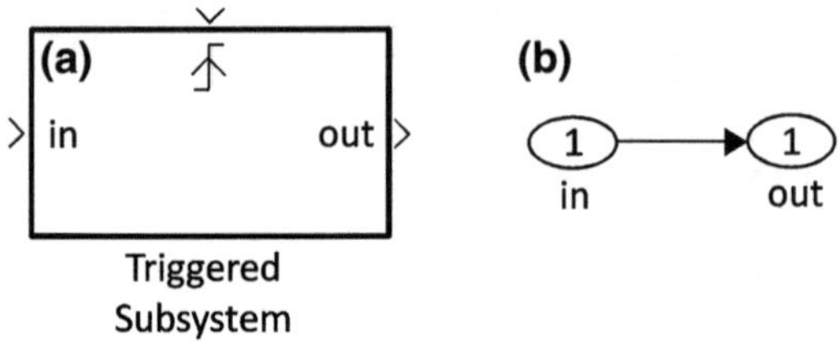

Fig. 6.18 **a** Triggered subsystem associated with **b** the model of sampler in Simulink

It is possible to include a number of Simulink blocks in a triggered subsystem. The Simulink blocks that are included in a triggered subsystem are executed only when an edge of the clock occurs. The example triggered subsystem in Fig. 6.17a is sensitive to the rising-edges of the signal Clock.

An Embedded Matlab Function is associated with a custom model that the user defines by means of a Matlab script. During a simulation, an Embedded Matlab Function is executed every time an event occurs. An Embedded Matlab Function that is included in a triggered subsystem is executed only on the edges of the clock of the triggered subsystem.

A Simulink triggered subsystem is executed only when an edge of its clock occurs. Consider a Simulink system that comprises a triggered subsystem that is clocked by an efficient oscillator model. The simulation of the system "triggered subsystem plus efficient oscillator" includes only the events that are scheduled by the oscillator model.

In the following sections, we illustrate how to model the remaining blocks of an ADPLL by means of triggered subsystems.

6.11 The Retiming Block

The retiming block is typically a sampler. Figure 6.18a shows a triggered subsystem that can be associated with the model in Fig. 6.18b to simulate a sampler in Simulink. The model in Fig. 6.18b reads the input, and simply assigns the read data to the output.

A sampler can also be modelled with an Embedded-Matlab-Function that includes the code in List. 6.9.

```
persistent mem; persistent clkPast; if
    isempty(clkPast)
     clkPast = 1;
end
if isempty(mem)
     mem = 1;
end
if (clk == 1) &&  (clkPast == 0) % Rising Edge
    Detected
mem = in; % Sampling the input signal
end
clkPast = clk;
out = mem;
```

Listing 6.9 Code to simulate a sampler with an Embedded-Matlab-Function in Simulink

6.12 Digital Filter

The output of a digital filter in the time domain can be represented by a finite difference equation [18]. The output $out[n]$ of an example digital filter with a pole and a zero is defined by:

$$out[n] = A_1 \cdot in[n] + A_2 \cdot in[n-1] + B \cdot out[n-1], \tag{6.14}$$

where $in[n]$ is the input of the filter, and the coefficients A_1, A_2 and B determine the positions of the pole z_p and zero z_z of the filter. The z-domain transfer function $H(z)$ that is associated with Eq. (6.14) is given by:

$$H(z) = A_1 \frac{z + z_z}{z - z_p}, \tag{6.15}$$

where $z_z = A_2/A_1$ and $z_z = B$.

Equation (6.14) can be implemented with a triggered subsystem that includes an Embedded Matlab Function. When Eq. (6.14) is realized with an Embedded Matlab Function, the discrete time index n increases by one every time the Embedded Matlab Function is executed. Moreover, in order to realize the terms $in[n-1]$ and $out[n-1]$ in Eq. (6.14), we need to access the values that the input and output signals of the Embedded Matlab Function had the last time it was executed. Only variables that are declared with the data type *persistent* maintain their values between consecutive executions of Embedded Matlab Functions [19]. The example Matlab script in List. 6.10 implements Eq. (6.14) by means of persistent variables.

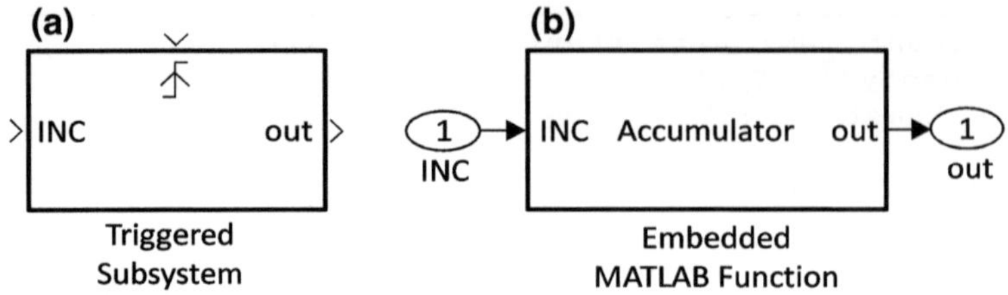

Fig. 6.19 **a** Triggered subsystem associated with **b** an the model an accumulator in Simulink

```
MemIn; persistent MemOut; if isempty(MemIn)
     MemIn = 0;
     MemOut = 0;
end
A1 = 0.005;
A2 = 0.005;
B = 0.9900;
out = A1*in + A2*MemIn + B*MemOut;
MemIn = in;
MemOut = out;
```

Listing 6.10 Matlab script for modeling a time discrete filter with an Embedded-Matlab-Function

6.13 Accumulator and Phase Sampler

Figure 6.19a shows the triggered subsystem associated with the Simulink model of an accumulator in Fig. 6.19b.

The code associated with the Embedded Matlab Function in Fig. 6.19b is shown in List. 6.11. The accumulator increments its output by INC on every rising edge of the signal that triggers the subsystem in Fig. 6.19a.

```
function out = Accumulator(INC)
persistent Mem;
if isempty(Mem)
     Mem = 0;
end
out = Mem + INC ;
Mem = out;
```

Listing 6.11 Matlab script for modeling an accumulator

The Simulink model of a phase sampler is shown in Fig. 6.20. The accumulator and the sampler in Fig. 6.20 are the models that we have shown in Figs. 6.18a and 6.19a, respectively.

The phase sampler model in Fig. 6.20 cannot be used in an ADPLL with the improved model of a DCO that has been described in Sect. 6.9. In fact, the phase

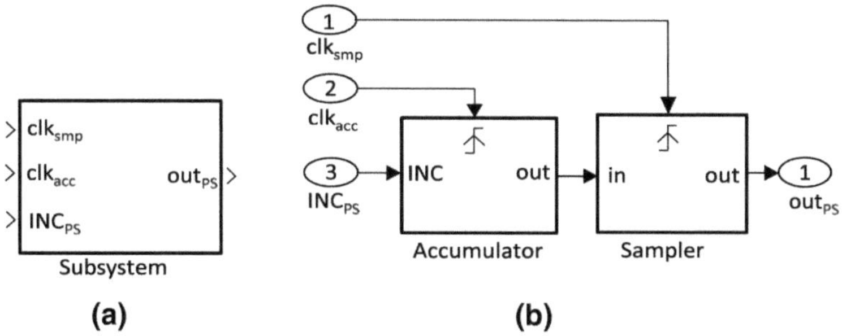

Fig. 6.20 **a** Subsystem associated with **b** the model of a phase sampler that is based on an accumulator and a sampler

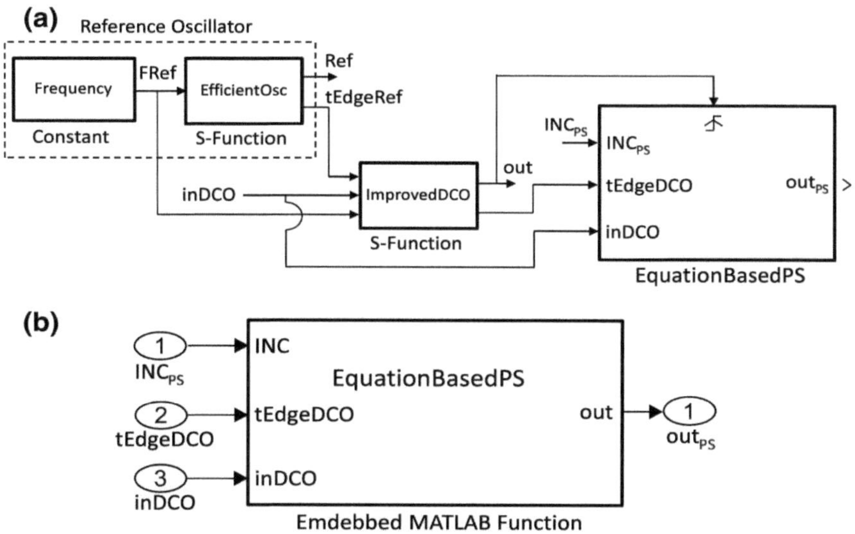

Fig. 6.21 Example block diagrams of **a** a Simulink model of the system "reference oscillator plus DCO and equation-based phase sampler", and **b** Embedded Matlab Function that is included in the triggered block labelled "EquationBasedPS" in Fig. 6.21a

sampler model in Fig. 6.20 needs every edge of the DCO to be simulated. However, the improved DCO model produces only a *subset* of the edges of a DCO.

An equation-based model of a phase sampler can be used in an ADPLL with the improved DCO. In fact, the output of a phase sampler in the time domain of the edges of the improved DCO can be calculated from INC_{PS}, the input $inDCO$ and output $tEdgesDCO$ of the improved DCO. An example equation-based model of a phase sampler implemented with a triggered subsystem is labelled "EquationBasedPS" in Fig. 6.21a. The Embedded-Matlab-Function that is included in the block "EquationBasedPS" in Fig. 6.21a is shown in Fig. 6.21b.

Listing 6.12 shows the Matlab script that is associated with the Embedded-Matlab-Function in Fig. 6.21b.

```
function out = PhaseSampler(INC,  tEdgesDCO, inDCO)
persistent mem;
persistent tEdgesDCOpast;
if isempty(mem)
    mem = 1;
end
if isempty(tEdgesDCOpast)
    tEdgesDCOpast = 0;
end
Kv = 1e9; % DCO gain
F0 = 100e6; % free-running frequency of the DCO
TDCO = 1/( Kv * inDCO + F0);
Nperiods = (tEdgesDCO - tEdgesDCOpast)/TDCO;
tEdgesDCOpast = tEdgesDCO;
mem = mem + INC*Nperiods;
out = mem;
```

Listing 6.12 Matlab script for modeling a phase sampler

In List. 6.12, the output of a phase sampler in the time domain of the edges of the improved DCO is obtained by accumulating INC_{PS} times the number $Nperiods$ of the DCO periods that are included between two consecutive edges of the improved DCO. $Nperiods$ is given by the duration of the time interval between two consecutive edges of the DCO divided by the duration of the DCO period. The duration of the time interval between two consecutive edges of the DCO is the difference between the current and past values of $tEdgesDCO$. The period $TDCO$ of the DCO is calculated from the input $inDCO$ of the improved DCO.

6.14 TDC

In an ADPLL, the TDC is used to measure the duration of the time interval $t_{in}[n]$ between the nth edge of the reference oscillator and the next significant edge of the DCO, as illustrated in Fig. 6.22.

We introduce the outputs $tEdgeRef$ and $tEdgeDCO$ in the efficient oscillator and improved efficient DCO models, respectively, in order to simplify the implementation of the model of a TDC. In fact, t_{in} is given by the difference between $tEdgeRef$ and $tEdgeDCO$, by definition. An example block diagram of a TDC model that uses $tEdgeRef$ and $tEdgeDCO$ is shown in Fig. 6.23a. The triggered subsystem labelled "TDC" in Fig. 6.23a includes the Embedded Matlab Function that is shown in 6.23b. The clock of the triggered subsystem labelled "TDC" in Fig. 6.23a is the output of the block "ImprovedDCO" because new values of $tEdgeRef$ and $tEdgeDCO$ are available as soon as a rising edge of the improved efficient DCO model occurs.

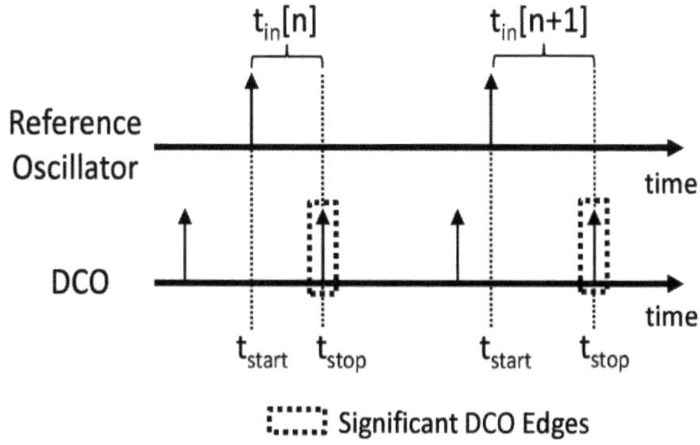

Fig. 6.22 Timing diagram of the rising edges of the reference oscillator and DCO with the duration $t_{in}[n]$ of the time interval that is measured by the TDC in an accumulator-based ADPLL

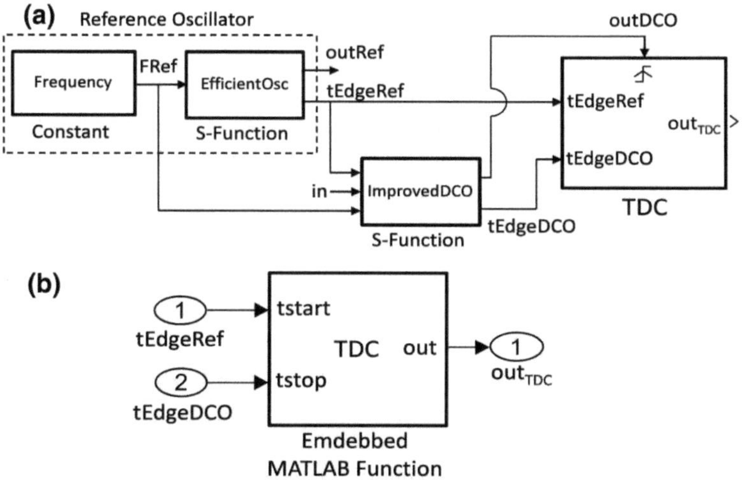

Fig. 6.23 Example block diagrams of **a** a Simulink model of the system "reference oscillator plus DCO and Time-to-Digital Converter", and **b** Embedded Matlab Function that is included in the triggered block labelled "TDC" in Fig. 6.23a

Listing 6.13 shows the Matlab code that can be included in the Embedded Matlab Function in Fig. 6.23b in order to model a TDC with a time resolution equal to res.

```
function out = TDC(tstart, tstop)
res = 1e-6;
out = res*floor((tstart-tstop)/res);
```

Listing 6.13 Matlab script for modeling the TDC

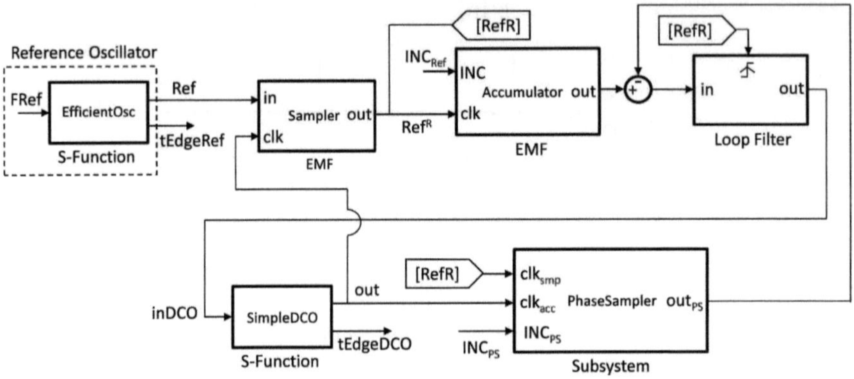

Fig. 6.24 Block diagram of a Simulink ADPLL model with a simple efficient DCO

6.15 Simulation Results

In this section, we compare two ADPLL models with a simple efficient DCO and an improved DCO, respectively. In order to show the benefits of the improved DCO with respect to a simple efficient DCO, we consider an ADPLL architecture with retiming and without a TDC. In fact, we will show that the presence of the TDCs does not affect the number of events produced by the ADPLL models. Figure 6.24 shows the block diagram of a Simulink ADPLL model that includes an efficient reference oscillator and a simple efficient DCO. The retiming block is implemented with a sampler.

Figure 6.25 illustrates the block diagram of an ADPLL model that includes an improved DCO. An equation-based phase sampler labelled "EquationBasedPS" has to be used to represent the phase sampler, as explained in Sect. 6.13. The output of the improved DCO is equal to the output of the retiming block, by the definition of a "significant edge" that was given in Sect. 6.6.

Figure 6.26 compares the simulated DCO frequencies of the models in Figs. 6.24 and 6.25 when INC_{Ref} is changed from 100 to 101. The step responses of the two models are identical. However, there is a huge difference between the times that are necessary to complete the simulations of the two ADPLL models. We used the Matlab Profiler to determine the completion times of the simulations on a computer with an Intel i7-Q720 and 4 GB of RAM.

The simulation of the ADPLL with the simple efficient DCO over the interval $[0, 3ms]$ took about 2 hours and 27 minutes. By contrast, the simulation of the ADPLL with the improved DCO was finished in less than 1 second.

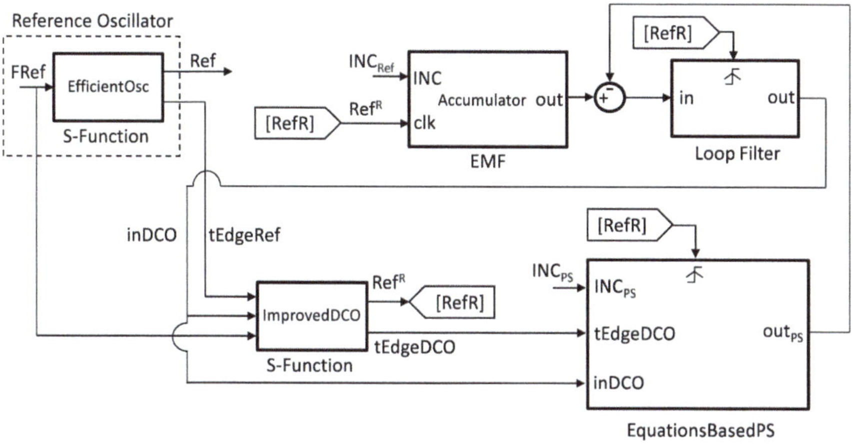

Fig. 6.25 Block diagram of a Simulink ADPLL model with an improved DCO

Fig. 6.26 Simulated DCO frequencies of the ADPLL models with simple efficient DCO and improved DCO when INC_{Ref} is changed from 100 to 101

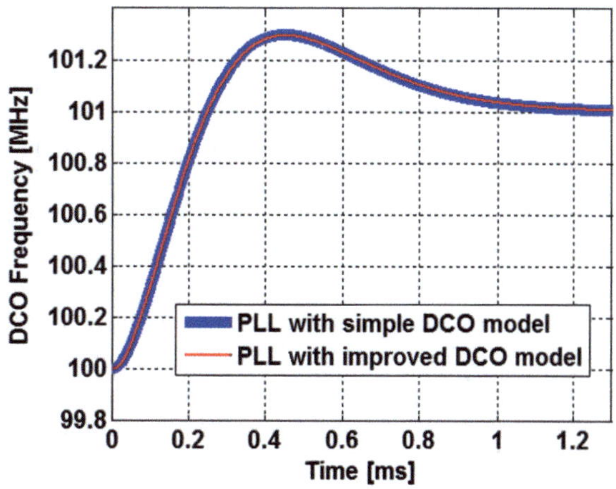

6.16 Increasing the Complexity of the DCO and the Phase Sampler

One of the main assumptions that we have made in this chapter is that the input of the DCO is constant between two consecutive significant edges of the DCO. This assumption is consistent with the fact that the ADPLL produces a measure of the phase error after a significant edge of the DCO has occurred. Hence, we assume that the digital filter that drives the DCO is clocked by a signal whose rising edges are the significant DCO's edges. However, the DCO input signal can change between two significant DCO's edges in ADPLLs with a DCO driven by a Digital Sigma-Delta Modulator [20, 21]. Furthermore, the DCO input can change because of a noise signal that adds to the output of the digital filter. The significant DCO's edges in the presence of a Digital Sigma-Delta Modulator or noise occur at instants $t_{eSDM}[n]$ or $t_{eNoise}[n]$, respectively, that are different from those $t_e[n]$ of an ADPLL without a Digital Sigma-Delta Modulator. The differences $t_{eSDM}[n] - t_e[n]$ and $t_{eNoise}[n] - t_e[n]$ can be calculated analytically when the histories of the output of the Digital Sigma-Delta Modulator or the noise signal are known, respectively. The DCO and the phase sampler models that we described in this chapter can be modified in order to take into account the effects of the Digital-Sigma Delta Modulator or noise with equations that calculate the differences $t_{eSDM}[n] - t_e[n]$ and $t_{eNoise}[n] - t_e[n]$. In order to keep our analysis simple, we haven't discussed this method in this chapter.

6.17 Conclusion

In this chapter, we discussed the behavioral modeling and simulation of accumulator-based ADPLLs. We gave a brief description of the main simulation issues for mixed-signal systems and ADPLLs. Then we showed how to implement an efficient behavioral ADPLL model that produces the minimum number of events during a simulation. We presented a simple efficient DCO model in Simulink that can control the simulation loop during a simulation. The simple efficient DCO model uses *user-defined* equations to manage the progress of the simulation by substituting the operation of the solver. We also showed an improved efficient DCO model that minimizes the number of events computed during the simulation of an ADPLL.

We showed how to realize behavioral models of the remaining building blocks in an ADPLL by means of Triggered Subsystems and Embedded Matlab Functions.

Finally, we compared the simulation results of efficient ADPLL models with a simple and an improved DCO. We showed that the step responses of the two models overlap exactly, even though the simulation of the ADPLL model with the improved DCO is hundreds of times faster than that of the ADPLL with the simple efficient DCO.

References

1. Mathworks, "Blocks - Sources - Clock," Simulink User's Guide. Available at: http://www.mathworks.co.uk/help/toolbox/simulink Accessed in Jun. 2012.
2. Accellera Organization, "Verilog-AMS Language Reference Manual," Version 2.3, Aug. 2008. Available at: http://www.accellera.org/apps/group_public/download.php/1388/VAMS-LRM-2-3.pdf Accessed in Jun. 2012.
3. Mathworks, "Choosing a Solver," Simulink User's Guide. Available at: http://www.mathworks.co.uk/help/toolbox/simulink/ug/f11-69449.html Accessed in Jun. 2012.
4. K. S. Kundert, O. Zinke, "The Designer's Guide To Verilog-AMS," Springer, 2004.
5. Mathworks, "Differential Equations in MATLAB," Tech-Note 1510. Available at: http://www.mathworks.co.uk/support/tech-notes/1500/1510.html Accessed in Jun. 2012.
6. Mathworks, "Interactive Matlab & Simulink Based Tutorials", available at: http://www.mathworks.co.uk/academia/student_center/tutorials/?s_cidglobal_nav Accessed in Jun. 2012.

7. Cadence Design Systems, "Virtuoso AMS Environment User Guide," Product Version 5.3, Apr. 2004.
8. Xilinx, "ISE In-Depth Tutorial," UG695 (v 12.1), Apr. 2009
9. Cadence Design Systems, "Virtuoso Spectre Circuit Simulator User Guide," Product Version 5.1.41, Jul. 2004.
10. Cadence Design Systems, "Cadence NC-Verilog Simulator Help," Product Version 3.4, Jan. 2002.
11. Cadence Design Systems, "Behaviour Modeling, Mixed-Signal And RF Simulation With Spectre Circuit Simulator," White Paper 2003. Downloadable at: http://w2.cadence.com/whitepapers/4640_Spectre_WPfnl.pdf Accessed in Jun. 2012.
12. Xilinx, "ISim User Guide," UG660 (v 13.1), Mar. 2011
13. Mentor Graphics Corporation, "ModelSim PE User's Manual," Software Version 10.0c, 2011
14. K. Kundert, H. Chang, D. Jefferies, G. Lamant, E. Malavasi, F. Sendig, "Design of Mixed-Signal Systems on Chip," Version 1a, 2006. Downloadable at: http://www.designers-guide.org/Design Accessed in Jun. 2012.
15. IEEE Computer Society, "Open SystemC Language Reference Manual" Downloadable at: http://standards.ieee.org/getieee/1666/download/1666-2011.pdf Accessed in Nov. 2012.
16. "PLL Design Assistant Program," a design tool developed by Mike Perrott and available at http://www.cppsim.com Accessed in Nov. 2012.
17. Mathworks, "Xilinx System Generator and HDL Coder," available at: http://www.mathworks.co.uk/fpga-design/Accessed in Sep. 2013.
18. S. Haykin, B. Van Veen, "Signals and Systems," Second Edition, *John Wiley & Sons Inc.*, 2003.
19. Mathworks, "Programming in Matlab/Variables and Functions in Memory/Persistent", Matlab User's Guide. Available at: http://www.mathworks.co.uk/help/techdoc/ref/persistent.html Accessed in Jun. 2012.
20. R. B. Staszewski, K. Waheed, S. Vemulapalli, F. Dulger, J. Wallberg, C.-M. Hung, O. Eliezer "Spur-Free All-Digital PLL in 65nm for Mobile Phones" *Solid-State Circuits Conference Digest of Technical Papers (ISSCC 2011)*, pp. 52–54, Feb. 2011.
21. N. Pavlovic, J. Bergervoet, "A 5.3GHz Digital-to-Time-Converter-Based Fractional-N All-Digital PLL" *Solid-State Circuits Conference Digest of Technical Papers (ISSCC 2011)*, pp. 54–56, Feb. 2011.

Chapter 7
Modelling and Estimating Phase Noise with Matlab

7.1 Introduction

The goal of this chapter is to illustrate how to model and estimate the phase noise of a sampled signal using Matlab. We first illustrate how the phase noise of a signal is related to the phase deviations of the signal. We then show how to model the phase noise of a noisy signal using Matlab. We relate the Power Spectral Density (PSD) of a noisy periodic signal analytically to that of its phase deviations. The PSD of the phase deviations is a measure of the phase noise of the noisy signal. The phase deviations of a signal can be estimated from the jitter of the signal. In fact, phase noise and jitter are representations of the noise of an oscillator in the frequency and in the time domains, respectively. We show how to derive equations that relate jitter and phase deviations. We illustrate how to estimate the phase deviations by interpolating the samples of a signal and calculating its jitter. We present Matlab scripts to compare the PSDs of the amplitude, actual and estimated phase deviations of a noisy signal. The Matlab script to estimate the phase deviations from the samples of a noisy signal can be used to extract the phase noise of signals from other simulation environments and also measurements.

7.1.1 Modelling Phase Noise

The output $osc_{ideal}(t)$ of an ideal noiseless oscillator can be defined by:

$$osc_{ideal}(t) = A\sin(\omega_0 t + \Phi_0), \tag{7.1}$$

where A, ω_0 and Φ_0 are the amplitude, angular frequency and phase of the sinusoid, respectively. When A, ω_0 and Φ_0 are constant, the Power Spectral Density of $osc_{ideal}(t)$ corresponds to a very narrow tone [1], as illustrated in Fig. 7.1. Figure 7.1 was produced by means of the Matlab script in List A.6 (see Appendix A.2). The

F. Brandonisio and M. P. Kennedy, *Noise-Shaping All-Digital Phase-Locked Loops*, Analog Circuits and Signal Processing, DOI: 10.1007/978-3-319-03659-5_7,

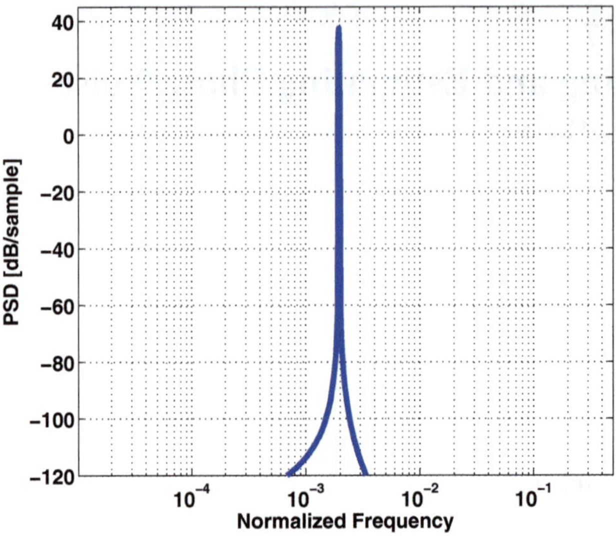

Fig. 7.1 Power spectral density of an example noiseless sinusoid

broadening of the peak in Fig. 7.1 is related to the windowing and averaging that are implemented by the function "pwelch" in List A.6.

The output of a noisy oscillator can be represented by [1–3]:

$$osc(t) = (A + \Delta A(t)) \sin(\omega_0 t + \Phi_0 + \Delta\Phi(t)), \tag{7.2}$$

where $\Delta A(t)$ and $\Delta\Phi(t)$ represent the deviations of the amplitude and phase of the noisy oscillator from the ideal noiseless amplitude and phase, respectively. In order to simplify our analysis, let us consider the case where $\Delta A(t)$ and Φ_0 are equal to zero.[1] When $\Delta A(t)$ and Φ_0 are zero, Eq. (7.1) simplifies to:

$$osc(t) = A \sin(\omega_0 t + \Delta\Phi(t)). \tag{7.3}$$

From Eq. (7.3), we can write:

$$osc(t) = A \sin(\omega_0 t) \cos(\Delta\Phi(t)) + A \cos(\omega_0 t) \sin(\Delta\Phi(t)). \tag{7.4}$$

When $\Delta\Phi(t)$ is much smaller than unity, Eq. (7.4) gives:

$$osc(t) \approx A \sin(\omega_0 t) + A \cos(\omega_0 t) \Delta\Phi(t). \tag{7.5}$$

[1] The effects of $\Delta\Phi(t)$ in an oscillator are more significant than those of $\Delta A(t)$ because an oscillator is an autonomous system that tends to correct any amplitude deviation $\Delta A(t)$ but *accumulates* the phase deviation $\Delta\Phi(t)$ [4].

We recall that multiplying a signal by a sinusoid is equivalent to shifting its spectrum by the frequency of the sinusoid. Hence, from Eq. (7.5), we obtain that the power spectral density $S_{osc}(\omega - \omega_0)$ of the output of a noisy oscillator is approximately given by:

$$\begin{aligned} S_{osc}(\omega - \omega_0) \approx {} & A^2 \left(\frac{\delta(\omega - \omega_0)}{4} + \frac{\delta(\omega + \omega_0)}{4} \right) \\ & + A^2 \left(\frac{S_{\Delta\Phi}(\omega - \omega_0)}{4} + \frac{S_{\Delta\Phi}(\omega + \omega_0)}{4} \right), \end{aligned} \tag{7.6}$$

where $\delta(\cdot)$ and $S_{\Delta\Phi}(\omega)$ are the Dirac delta and the Power Spectral Density of the phase deviations $\Delta\Phi(t)$, respectively. Equation (7.6) is associated with a *double sided* PSD. Some PSD algorithms such as the Matlab functions "periodogram" and "pwelch" return the *single sided* PSD of their argument when the argument is real [5]. Using Eq. (7.6) one can also write the single-sided form of $S_{osc}(\omega)$, which is given by:

$$S_{osc}(\omega - \omega_0) \approx A^2 \left(\frac{\delta(\omega - \omega_0)}{2} \right) + A^2 \left(\frac{S_{\Delta\Phi}(\omega - \omega_0)}{2} \right). \tag{7.7}$$

Measurement instruments and simulators usually return $L(\Delta\omega)$ instead of $S_{osc}(\omega - \omega_0)$ [1, 4]. $L(\Delta\omega)$ is defined by [1, 4]:

$$L(\Delta\omega) = \frac{2 S_{osc}(\omega)}{A^2}, \tag{7.8}$$

where $\Delta\omega = \omega - \omega_0$. As can be seen from Eq. (7.8), $L(\omega)$ is equal to $S_{osc}(\Delta\omega)$ normalized to the power of the tone that is associated with $\delta(\omega - \omega_0)$. By substituting Eq. (7.7) into (7.8), we obtain:

$$L(\Delta\omega) \approx \delta(\Delta\omega) + PSD_{\Delta\Phi}(\Delta\omega). \tag{7.9}$$

In the frequency domain, $L(\Delta\omega)$ is a common measure of the phase noise [4]. When $\Delta\Phi(t)$ is known for all time t, Eq. (7.3) can be used to realize an analytical model of an oscillator with phase noise. In order to illustrate an example model of an oscillator in a *simulator*, we assume that only samples $\Delta\Phi(t[n])$ of $\Delta\Phi(t)$ at times $t[n]$ are known. The index n is associated with the discrete time domain of the simulation. Hence, the output $osc(t[n])$ of an oscillator model in a simulation is determined by:

$$osc(t[n]) = A \sin(\omega_0 t[n] + \Delta\Phi(t[n])). \tag{7.10}$$

So far we, haven't specified an expression for $\Delta\Phi(t[n])$. In order to derive an expression for $\Delta\Phi[n]$, we note that an oscillator is an autonomous system in which phase deviations accumulate over time. However, the inherent accumulation of phase deviations does not affect the noise contributions of external sources that are added directly to the output of the oscillator, as shown in [4]. Hence, the phase of the output signal of an oscillator can include a noise component which is *not* accumulated.

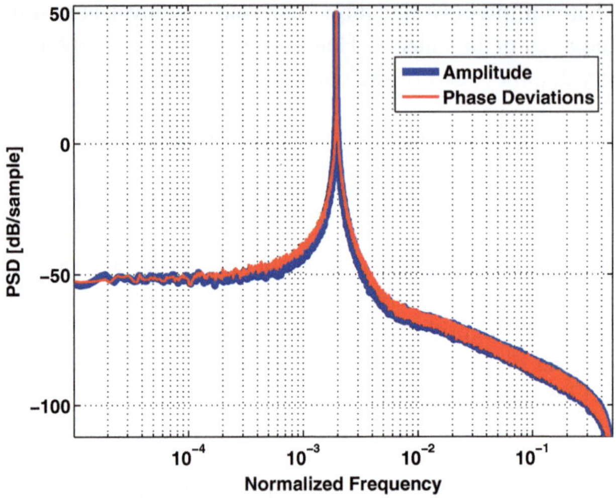

Fig. 7.2 PSDs of the output and phase deviations of a noisy sinusoid

This noise component can be considered synchronous with the output of the oscillator. Accumulating and synchronous phase deviations are related to accumulating and synchronous jitter, as will be shown in Sect. 7.1.2. In order to take account of both the accumulating and synchronous phase deviations, we assume that $\Delta\Phi(t[n])$ is of the form:

$$\Delta\Phi(t[n]) = \Phi_{synch}(t[n]) + \sum_{k=1}^{n} \Phi_{acc}(t[k]), \tag{7.11}$$

where $\Phi_{synch}(t[n])$ and $\Phi_{acc}(t[k])$ represent the accumulating and synchronous components of the phase deviations, respectively. The Matlab code in List A.4 presents an example oscillator model that is based on Eqs. (7.10) and (7.11). In List A.4, $\Phi_{synch}(t[n])$ and $\Phi_{acc}(t[k])$ are random variables with Gaussian distributions.

Figure 7.2 compares $L(\Delta\omega)$ and $PSD_{\Delta\Phi}(\Delta\omega)$ for the signal at the output of the oscillator model and $\Delta\Phi(t[n])$ that are produced by List A.4. The x-axis in Fig. 7.2 is normalized to the sampling frequency. Figure 7.2 has been produced by means of List A.5. The two PSDs of the amplitude and phase deviations of the noisy signal are almost identical, as expected from Eq. (7.9).

7.1.2 Modelling Jitter

Jitter is an undesired fluctuation in the timing of events [6]. Jitter can be seen as a manifestation of phase noise in the time domain. In order to define the jitter of the output signal of an oscillator, we assume that one of the possible values of the signal is a reference level. The output of a noisy oscillator with period T_0 crosses the reference level in the same direction at the instants $t_e[i]$, where i = 0, 1, 2,..., that are not necessarily spaced by T_0. By contrast, the output of an ideal noiseless oscillator with period T_0 crosses the reference level in a particular direction at the instants $t_{eideal}[i]$, where i = 0, 1, 2,..., that are equally spaced by T_0. The jitter of the output signal of the noisy oscillator can be measured by means of the difference $\Delta T[i]$, where $\Delta T[i]$ is given by [4]:

$$\Delta T[i] = t_e[i] - t_{eideal}[i], \tag{7.12}$$

and i denotes the ith period of the signal. Jitter can be separated into two main types: (i) accumulating and (ii) synchronous, which can be associated with the accumulating and synchronous phase deviations that we mentioned in Sect. 7.1.1. Accumulating jitter is typical of autonomous systems, such as free-running, voltage- and digitally-controlled oscillators. In fact, autonomous systems do not have a time reference and tend to accumulate phase deviations [6].

Driven circuits, such as dividers and phase frequency detectors, are characterized by synchronous jitter. In order to relate jitter to phase noise, let us consider a sampled sinusoid $osc[n]$, where $osc[n]$ is given by:

$$osc(t[n]) = A\sin(\omega_0 t[n] + \Delta\Phi(t[n])), \tag{7.13}$$

where $t[n]$ are the sampling instants and $\Delta\Phi(t[n])$ is the phase deviation associated with the sample at the instant $t[n]$. Assume that the reference level for determining the crossings is zero. The crossings are determined on the rising edges of the signal. Hence, from Eq. (7.19) we have that $t_{eideal}[i]$ is given by:

$$t_{eideal}[i] = \left(\frac{2\pi}{\omega_0}\right) i, \tag{7.14}$$

where the index i denotes the ith period. Assume that $t_{prv}[i]$ and $t_{nxt}[i]$ are the sampling instants that are the closest to $t_{eideal}[i]$ and that the interval $[t_{prv}[i], t_{nxt}[i]]$ includes $t_{eideal}[i]$.

The instants $t_e[i]$ can be determined approximately by interpolating linearly between the samples that are associated with $t_{prv}[i]$ and $t_{nxt}[i]$, as shown in Fig. 7.3. $t_{eideal}[i]$ can be calculated by means of Eq. (7.14). $\Delta T[i]$ is determined when $t_e[i]$ and $t_{eideal}[i]$ are known. Hereafter, we derive an expression that relates $\Delta\Phi(t_{prv}[n])$ to $\Delta T[i]$. The relation that we will derive will show how jitter and phase deviations are related.

Consider a noiseless and a noisy sinusoid that are defined by:

$$osc_{ideal}(t) = A\sin(\omega_0 t), \tag{7.15}$$

and

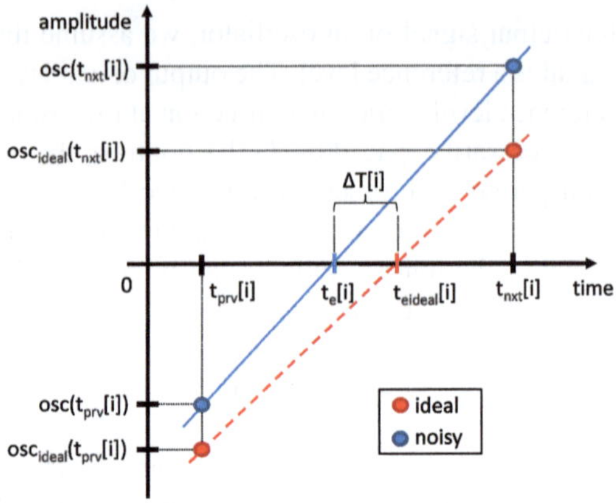

Fig. 7.3 Example linear interpolation between samples of a noisy and a noiseless sinusoid

$$osc(t) = A\sin(\omega_0 t + \Delta\Phi(t)), \tag{7.16}$$

respectively. The phase deviation $\Delta\Phi(t)$ in Eq. (7.16) determines the misalignment between the crossing instants of the noisy and noiseless signals. Assume that $\Delta\Phi(t)$ is known. Moreover, assume that the reference level that is used to determine the crossings of the ideal and the noisy sinusoids is equal to zero. Crossings are calculated only when the slopes of the signals are positive. From Eq. (7.18), we deduce that the crossings of the ideal sinusoid are given by:

$$t_{eideal}[i] = (2\pi/\omega_0)i, \tag{7.17}$$

where the index i denotes the ith period. Equation (7.16) can be used to determine the crossing instants $t_e[i]$ for the noisy signal in the continuous time domain, as shown in [6]. Then, $t_e[i]$ and $t_{eideal}[i]$ can be used to determine the jitter.

However, in practical cases and in simulation, only a *sampled* version of a noisy signal is available. Hence, it is convenient to determine the jitter by assuming a discrete time domain associated with the index n of the samples of the signals. In the discrete time domain with samples indexed by n, Eq. (7.16) can be written as:

$$osc_{ideal}(t[n]) = A\sin(\omega_0 t[n]) \tag{7.18}$$

and

$$osc(t[n]) = A\sin(\omega_0 t[n] + \Delta\Phi(t[n])), \tag{7.19}$$

where $t[n]$ are the sampling instants. Notice that we assume that the signals osc_{ideal} and osc are sampled at the same time instants in order to define a relationship between the samples of the noisy signal and $\Delta\Phi(t[n])$. Moreover, we assume that the time instants $t[n]$ are equally spaced by the sampling time T_{step}. The ith periods of the signals $osc_{ideal}(t[n])$ and $osc(t[n])$ are characterized by two sampling instants $t_{prv}[i]$ and $t_{nxt}[i]$ that are the closest to $t_{eideal}[i]$, where the interval $[t_{prv}[i], t_{nxt}[i]]$ includes

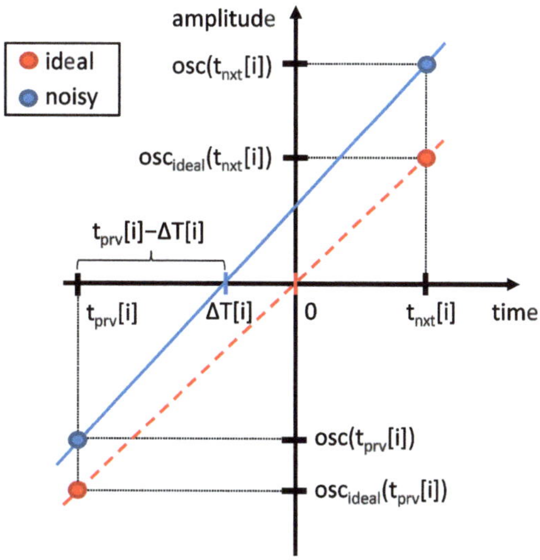

Fig. 7.4 Reference system and notation that we assume in this work to describe the crossings of noisy and noiseless sampled sinusoids

$t_{eideal}[i]$. In order to simplify the calculations, we shift the time axis by $t_{eideal}[i]$. Hence, the origin of the reference system is associated with $t_{eideal}[i]$. A measure of the jitter of a signal is the difference $\Delta T[i]$ [4], where $\Delta T[i]$ is given by:

$$\Delta T[i] = t_e[i] - t_{eideal}[i]. \tag{7.20}$$

Figure 7.4 shows an example of the notation and the reference system that we use. As illustrated in Fig. 7.4, it is possible to estimate the value of $\Delta T[i]$ by means of a linear interpolation between the samples of the noisy signal at the instants $t_{prv}[i]$ and $t_{nxt}[i]$. In order to keep the notation simple, we refer to the *estimated* value of $\Delta T[i]$ as $\Delta T[i]$. Thus,

$$\frac{t_{prv}[i] - \Delta T[i]}{osc(t_{prv}[i])} = \frac{t_{nxt}[i] - t_{prv}[i]}{osc(t_{nxt}[i]) - osc(t_{prv}[i])}, \tag{7.21}$$

From Eq. (7.21), we can write:

$$\Delta T[i] = -\frac{\left(t_{nxt}[i] - t_{prv}[i]\right) osc(t_{prv}[i])}{osc(t_{nxt}[i]) - osc(t_{prv}[i])} + t_{prv}[i]. \tag{7.22}$$

Equation (7.21) can be used to calculate $\Delta T[i]$ from the samples of a noisy oscillator. In order to relate $\Delta T[i]$ to the phase deviations $\Delta\Phi(t[n])$ of a noisy oscillator, notice that $t[n]\omega_0$ is close to an integer multiple of 2π when $t[n]$ is very close to $t_{eideal}[i]$, i.e. when $t_{nxt}[i]$ and $t_{prv}[i]$ are very close to $t_{eideal}[i]$. Hence, when $t_{nxt}[i]$ and $t_{prv}[i]$ are very close to $t_{eideal}[i]$, Eq. (7.19) can be written as:

$$osc(t[n]) \approx A(\omega_0 t[n] + \Delta\Phi(t[n])). \tag{7.23}$$

By substituting Eq. (7.23) into (7.21) we obtain a relationship between $\Delta T[i]$, $\Phi(t_{nxt}[n])$ and $\Phi(t_{prv}[n])$, where $\Phi(t_{nxt}[n])$ and $\Phi(t_{prv}[n])$ are two unknown variables. This problem can be solved numerically by considering another pair of samples close to $t_{eideal}[i]$. However, we can obtain a single approximate equation in one unknown variable by assuming that the slope of the solid line in Fig. 7.4 is approximately equal to that of the noiseless signal, which is equal to ω_0. Hence, we assume:

$$\frac{t_{nxt}[i] - t_{prv}[i]}{osc(t_{nxt}[i]) - osc(t_{prv}[i])} \approx \omega_0. \tag{7.24}$$

Thus, by substituting Eq. (7.24) into (7.21) and expressing $\Delta T[i]$, we have:

$$\Delta T[i] \approx -\frac{osc(t_{prv}[i])}{\omega_0} + t_{prv}[i]. \tag{7.25}$$

By substituting Eq. (7.23) into (7.25), we obtain:

$$\Delta T[i] \approx -\frac{\Delta\Phi(t_{prv}[i])}{\omega_0}. \tag{7.26}$$

Finally, and by expressing $\Delta\Phi(t_prv[i])$ from Eq. (7.26), we have:

$$\Delta\Phi(t_{prv}[i]) \approx -\omega_0 \Delta T[i]. \tag{7.27}$$

As can be derived from Eq. (7.27), the difference between the ideal and actual transition $\Delta T[i]$ of the ith period is approximately proportional to the phase deviation $\Delta\Phi(t_{prv}[i])$ that is associated with the sample of the noisy signal at the time instant $t_{prv}[i]$.

When the instant $t_{prv}[i]$ is very close to $t_{eideal}[i]$, the phase deviation $\Delta\Phi(t_{prv}[i])$ is approximately equal to $\Delta\Phi(t[n])$. Moreover, $t_{prv}[i+1] - t_{prv}[i]$ is approximately equal to the period $T_0 = 2\pi/\omega_0$ of the noiseless signal. Hence, $\Delta\Phi(t_{prv}[i])$ is approximately equal to the phase deviations $\Delta\Phi(t)$ sampled at the frequency ω_0. Figure 7.5 shows a comparison between the actual and extracted phase deviation $\Delta\Phi(t[n])$ and $\Delta\Phi(t_{prv}[n])$, respectively. Figure 7.5 was produced by means of List A.7. The script in List A.7 finds $t_{prv}[i]$ and $t_{nxt}[i]$, uses linear interpolation to calculate $\Delta T[i]$, and calculates $\Delta\Phi(t_{prv}[i])$ by means of Eq. (7.27). The noisy signal that was processed to produce Fig. 7.5 was generated by means of List A.4. As can be seen from Fig. 7.5, the actual and extracted phase deviations match.

The PSD of the phase deviations estimated by means of List A.7 is approximately equal to the PSD of the amplitude of the noisy signal in the frequency offset range $[-\omega_0/2, \omega_0/2]$ from the main peak of the noisy signal, where ω_0 is the angular frequency of the noiseless signal. In order to simplify our analysis, let us assume that the samples of the noisy sinusoid are equally spaced with a sampling period

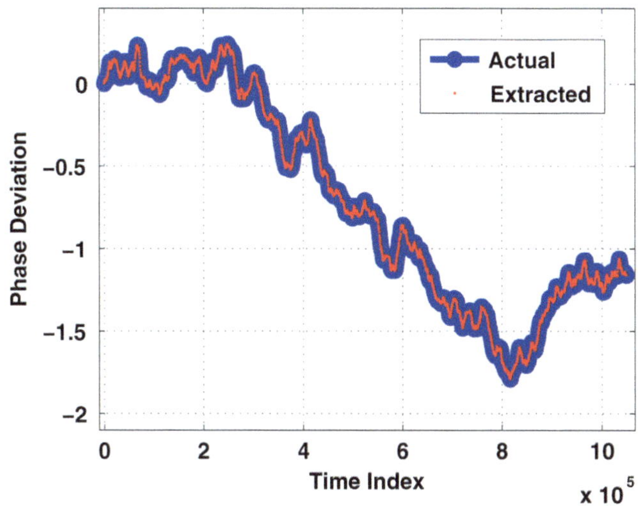

Fig. 7.5 Actual and extracted phase deviations of an example noisy sinusoid produced using Matlab

equal to $2\pi/\omega_s$, where ω_s is the sampling frequency. The phase noise of the signal estimated from $\Delta\Phi(t[n])$ is defined over the frequency offset $[-\omega_s/2, \omega_s/2]$ around the main peak of the noisy signal. Figure 7.6 compares the PSDs of $\Delta\Phi(t[n])$ and $\Delta\Phi(t_{prv}[i])$, where $\Delta\Phi(t[n])$ represents the actual phase deviations of a noisy sinusoid and $\Delta\Phi(t_{prv}[i])$ is calculated from the crossing instants $t_{eideal}[i]$ and $t_e[i]$. The spectra in Fig. 7.6 are plotted only for frequencies that are higher than that of the main peak of the noisy signal because the rest of the spectrum is symmetric with respect to the main peak. Furthermore, the x-axis is normalized to ω_s. Figure 7.6 was produced by means of the Matlab script in List A.8. The noisy sinusoid and the phase deviation signal $\Delta\Phi(t_{prv}[i])$ were generated by means of Lists A.4 and A.7, respectively.

Using estimated phase deviations, it is possible to obtain approximately the PSD of the actual phase deviations of the noisy signal over a limited range of normalized frequencies which is given by $[-\omega_0/\omega_s, \omega_0/\omega_s]$, as shown in Fig. 7.6.

7.1.3 Normalizations

Phase noise is often expressed in dBc/Hz. In order to express the PSDs that we have reported in Figs. 7.2 and 7.6 in dBc, one should divide the PSDs by the power of the main peak in the spectrum of the noisy sinusoid. Then, dividing the power spectral density in dBc by the frequency ω_s of the noiseless signal gives the power spectral density in dBc/Hz.

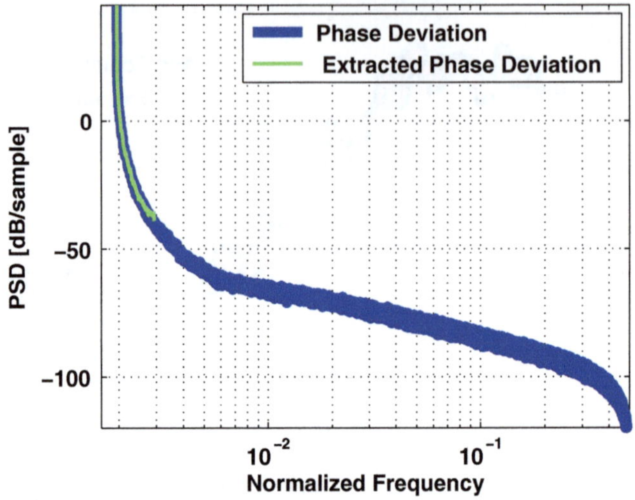

Fig. 7.6 Actual and extracted phase deviations of an example noisy sinusoid produced in Matlab

7.2 Conclusion

In this chapter we have derived equations to relate the jitter, the phase deviations and the phase noise of a signal. We have shown how to model a noisy oscillator in Matlab and we have presented Matlab scripts to compare the PSDs of the amplitude, actual and estimated phase deviations of a noisy signal. The Matlab scripts and the analytical predictions that we described in this chapter can be used to calculate the phase noise of signals from measurements and other simulation environments.

References

1. A. Abidi, "Basics of Phase Noise and Jitter", University of California, Los Angeles Electrical Engineering Department. Downloadable at http://ewh.ieee.org/r7/toronto/chapters/ssc/JitterXandXPhaseXNoise.pdf. Last access in October 2012
2. Agilent "Agilent E5505A Phase Noise Measurement System, User's Guide", Agilent Technologies Inc., 2004–2012.
3. Maxim "Clock (CLK) Jitter and Phase Noise Conversion", Application Note 3359, Dec. 2004.
4. K. Kundert, "Predicting the Phase Noise and Jitter of PLL-Based Frequency Synthesizers", Designer's Guide Consulting Inc., downloadable at http://www.designers-guide.org/analysis/PLLnoise+jitter.pdf. Last access in October 2012
5. Matworks, "Periodogram, PSD using periodogram", Matlab 2010 User's Guide.
6. J. Phillips, K. Kundert "Noise in Mixers, Oscillators, Samplers, and Logic An Introduction to Cyclostationary Noise", Designer's Guide Consulting Inc., downloadable at http://www.designers-guide.org/Theory/cyclo-paper.pdf. Last access in October 2012

Appendix A

The Matlab scripts shown in this chapter are downloadable at http://cas.tyndall.ie/.

A.1 Matlab Scripts for Simulating ADPLLs

```
%%%%%%%%%%%%%%%%%%%%%%%%%%%%%%%%%%%%%
%%%    PLL MODEL WHICH INCLUDES   %%%%
%%%     A TDC AND A DCO WITH      %%%%
%%%       MACHINE PRECISION       %%%%
%%%         RESOLUTIONS           %%%%
%%%%%%%%%%%%%%%%%%%%%%%%%%%%%%%%%%%%%
clc
close all
clear all
%%%%%%%%%%%%%%%%%%%%%%%%%%%%%%%%%%%%%
%%% Parameters obtained from   %%%%%
%%%%%  PLL Assistant Designer  %%%%%
%%%%%%%%%%%%%%%%%%%%%%%%%%%%%%%%%%%%%
% fz/fo = 0.22
% order 2
% shape Butter
% Ref freq 10e6
% out freq = 1e9
% fz = 2.2e5;
% wz = 2*pi*fz;
% fp = 1.734e5;
% wp = 2*pi*fp;
% N = 100;
% Kccpsim = 7.272e10;
%%%%%%%%%%%%%%%%%%%%%%%%%%%%%%%%%%%%%%
Nsamples = 2^22;
fz = 2.2e4;
wz = 2*pi*fz;
```

F. Brandonisio and M. P. Kennedy, *Noise-Shaping All-Digital Phase-Locked Loops*, Analog Circuits and Signal Processing, DOI: 10.1007/978-3-319-03659-5,

```
fp = 1.734e5;
wp =2*pi*fp;
fref = 10e6;
N = 100;
Kccpsim = 7.272e10;
KccpsimAdapted = Kccpsim/wz;
kv = 10e6;
kp = 1;
klf = N*KccpsimAdapted/(kv*kp);
%%%%%%%%%%%%%%%%%%%%%%%%%
%%%  Inizialization   %%%
%%%%%%%%%%%%%%%%%%%%%%%%%
freq_freerunning = fref*N; % freerunning frequency of
   the controlled oscillator
output_phase_dco = zeros(Nsamples,1);
phase_input = zeros(Nsamples,1);
phase_err = zeros(Nsamples,1);
freq_dco = zeros(Nsamples,1);
out = zeros(Nsamples,1);
outExtra = zeros(Nsamples,1);
outExtraGain = zeros(Nsamples,1);
input_dco = zeros(Nsamples,1);
output_phase_dco(1) = 1;
freq_dco(1) = freq_freerunning;
eq = zeros(Nsamples,1);
tin = zeros(Nsamples,1);
outTDC = zeros(Nsamples,1);
outTDCwithGain = zeros(Nsamples,1);
eqDeltaC = zeros(Nsamples,1);
Finput = fref * ones(Nsamples,1); % Input Frequency
time_array =(1/fref) * (0:1: Nsamples-1); % Time Array
fs = fref; % Sampling Frequency of the Digital Filter
%%%%%%%%%%%%%%%%%%%%%%%%%%%%%%%%%%%%%%%%%%%%%%%%%%%%%%%%
%%%%%%%  FILTER DESIGN: ZERO plus INTEGRATOR   %%%%%%%%%
%%%%%%%%%%%%%%%%%%%%%%%%%%%%%%%%%%%%%%%%%%%%%%%%%%%%%%%%
syms z z1 outz inz p1;
s = 2 * fs * (z-1)/(z+1);
Hz = (s + z1) / (s);
Hz = simple(Hz);
[N,D] = numden(Hz);
Cnum = coeffs(N, z);
Cden = coeffs(D, z);
Cnum1 = Cnum(1);
Cnum2 = Cnum(2);
Cden1 = Cden(1);
Cden2 = Cden(2);
fsval = 2*pi*fs; % sampling frequency in the discrete
   time domain
z1val =  wz;
Cnum1val = double(subs(Cnum1, {z1, fs},{z1val,
   fsval}));
Cnum2val = double(subs(Cnum2, {z1, fs},{z1val,
```

```
    fsval}));
Cden1val = double(subs(Cden1, {z1, fs},{z1val,
    fsval}));
Cden2val = double(subs(Cden2, {z1, fs},{z1val,
    fsval}));
A = (Cnum2val/ Cden2val);
B = (Cnum1val/ Cden2val);
C =  - (Cden1val/ Cden2val);
%%%%%%%%%%%%%%%%%%%%%%%%%%%%%%%%%%%%%%%%%%%%%%%%
%%%%%%%  FILTER DESIGN: EXTRA POLE  %%%%%%%%%
%%%%%%%%%%%%%%%%%%%%%%%%%%%%%%%%%%%%%%%%%%%%%%%%
s = 2 * fs * (z-1)/(z+1);
Hz = p1/(s+p1);
[N,D] = numden(Hz);
Cnum = coeffs(N, z);
Cden = coeffs(D, z);
Cnum1 = Cnum(1);
Cnum2 = Cnum(2);
Cden1 = Cden(1);
Cden2 = Cden(2);
p1val = wp;
fsval = 2*pi*fs; % sampling frequency in the discrete
    time domain
Cnum1valextra = double(subs(Cnum1, {p1, fs},{p1val,
    fsval}));
Cnum2valextra = double(subs(Cnum2, {p1, fs},{p1val,
    fsval}));
Cden1valextra = double(subs(Cden1, {p1, fs},{p1val,
    fsval}));
Cden2valextra = double(subs(Cden2, {p1, fs},{p1val,
    fsval}));
Aextra = (Cnum2valextra/ Cden2valextra);
Bextra = (Cnum1valextra/ Cden2valextra);
Cextra =  - (Cden1valextra/ Cden2valextra);
%%%%%%%%%%%%%%%%%%%%%%%%%%%%%%%%%%%%%%%%%%%%%%%%
%%%%%%%  FILTER DESIGN: INTEGRATOR  %%%%%%%%%
%%%%%%%%%%%%%%%%%%%%%%%%%%%%%%%%%%%%%%%%%%%%%%%%
s = 2 * fs * (z-1)/(z+1);
Hz = 1/s;
[N,D] = numden(Hz);
Cnum = coeffs(N, z);
Cden = coeffs(D, z);
Cnum1 = Cnum(1);
Cnum2 = Cnum(2);
Cden1 = Cden(1);
Cden2 = Cden(2);
fsval = 2*pi*fs; % sampling frequency in the discrete
    time domain
Cnum1valint = double(subs(Cnum1,  fs, fsval));
Cnum2valint = double(subs(Cnum2,  fs, fsval));
Cden1valint = double(subs(Cden1,  fs, fsval));
Cden2valint = double(subs(Cden2,  fs, fsval));
```

```
Aint = (Cnum2valint/ Cden2valint);
Bint = (Cnum1valint/ Cden2valint);
Cint =  - (Cden1valint/ Cden2valint);
%%%%%%%%%%%%%%%%%%%%%%%%%%%%%%%%%%%%%%%%%%%%%%%%%%%%%%
%%%%%% Noise to be added to the input of the DCO %%%%%%
%%%%%%%%%%%%%%%%%%%%%%%%%%%%%%%%%%%%%%%%%%%%%%%%%%%%%%
foff = 1e6; % [Hz]
noise_at_foff = -130; % [dBc/Hz]
noise_var = (foff^2)/(kv^2)*10^(noise_at_foff/10);
Ts = 1/fref;
AdditiveNoiseDCOinput =
     sqrt(noise_var/Ts)*randn(1,length(time_array));
VarAdditiveNoiseDCOinput = var(AdditiveNoiseDCOinput);

for n = 2:1:length(time_array)
%%%%%%%%%%%%%%%%%%%%%%%%%%%%%%%%%%%%
%%%   Freq to Phase conversion    %%%
%%%       of the input Fref       %%%
%%%%%%%%%%%%%%%%%%%%%%%%%%%%%%%%%%%%
phase_input(n) =  (Cnum2valint/ Cden2valint) *
   Finput(n)  + (Cnum1valint/ Cden2valint) *
   Finput(n-1) - (Cden1valint/ Cden2valint) *
   phase_input(n-1) ;
if n == 2
   N = 100;
end
%%%%%%%%%%%%%%
%%%   TDC   %%%
%%%%%%%%%%%%%%
phase_err(n) = (phase_input(n) - output_phase_dco(n
   -1));
tin(n) = (1/fref)* phase_err(n);
resTDC = 1e-19; % [seconds] rest = 1e-19 introduces
   negligible quant error
outTDC(n) = resTDC*floor(tin(n)/resTDC);
eq(n) = resTDC* (tin(n)-floor(tin(n)));
outTDCwithGain(n) = kp* outTDC(n)*fref;
%%%%%%%%%%%%%%%%%%%%%%%%%%%%%%%%%%%%
%%%  FILTER: zero plus integrator  %%%
%%%%%%%%%%%%%%%%%%%%%%%%%%%%%%%%%%%%
out(n)  =   Cnum2val/Cden2val  * outTDCwithGain(n) +
   Cnum1val/  Cden2val * outTDCwithGain(n-1) -
   Cden1val/ Cden2val    * out(n-1);
%%%%%%%%%%%%%%%%%%%%%%%%%%%%%%%%%%%%
%%%        FILTER: Extra Pole     %%%
%%%%%%%%%%%%%%%%%%%%%%%%%%%%%%%%%%%%
outExtra(n) =   (Cnum2valextra/ Cden2valextra) *
   out(n)  +  (Cnum1valextra/ Cden2valextra) * out(n-1)
   - (Cden1valextra/ Cden2valextra) * outExtra(n-1) ;
%%%%%%%%%%%%%%%%%%%%%%%%%%%%%%%%%%%%%%%%%%%%
%%%%           Digital Filter Gain          %%%%%
%%%%%%%%%%%%%%%%%%%%%%%%%%%%%%%%%%%%%%%%%%%%
```

```
outExtraGain(n) = klf * outExtra(n);
input_dco(n) = outExtraGain(n); % +
    AdditiveNoiseDCOinput(n);
%%%%%%%%%%%%%%%%%%%%%%%%%%%%%%%%%%%%%%%%%%%%%%%%
%%%%          DCO with quantization         %%%%%
%%%%%%%%%%%%%%%%%%%%%%%%%%%%%%%%%%%%%%%%%%%%%%%%
c = 10e-12;
l = 100e-9;
kdcoequivalent = -1/(2*sqrt((l*c)^3));
deltaC = kv*(1/kdcoequivalent) * input_dco(n);
resDCO = 1e-20; % [Farads] 1e-20 F is negligible
deltaCq = resDCO * floor(deltaC/resDCO);
eqDeltaC(n) = resDCO*(deltaC/resDCO -
    floor(deltaC/resDCO));
freq_dco(n) = 1/sqrt(l*c) + kdcoequivalent*deltaCq +
   kv*AdditiveNoiseDCOinput(n);
output_phase_dco(n) = (Cnum2valint/ Cden2valint) *
   freq_dco(n)/N + (Cnum1valint/ Cden2valint) *
   freq_dco(n-1)/N - (Cden1valint/ Cden2valint) *
   output_phase_dco(n-1);
end
%%%%%%%%%%%%%%%%%%%%%%%%%%%%%%%%%%%%%%%%%%%%%%%%%%
%%%%% Selection of the steady state samples   %%%%%
%%%%%%%%%%%%%%%%%%%%%%%%%%%%%%%%%%%%%%%%%%%%%%%%%%
SteadyStateOffset = 1e5;
freq_dco_steady = freq_dco(SteadyStateOffset:end);
%%%%%%%%%%%%%%%%%%%%%%%%%%%%%%%%%%%%%%%%%%%%%%%%%%
%%%%%    Calculate the DCO's Phase Noise      %%%%%
%%%%%%%%%%%%%%%%%%%%%%%%%%%%%%%%%%%%%%%%%%%%%%%%%%
num_segments = 16;
DCOin =  freq_dco_steady/kv;   %; input_dco;
phase = filter(Ts*2*pi*kv,[1 -1],DCOin-mean(DCOin));
window_length = floor(length(phase)/num_segments);
[PSDphase,f] =
   pwelch(phase,window_length,[],[],1/Ts,'twosided');
%%%%%%%%%%%%%%%%%%%%%%%%%%%%%%%%%%%%%%%%%%%%%%%%%%
%%%%%          Analytical Preditctions        %%%%%
%%%%%%%%%%%%%%%%%%%%%%%%%%%%%%%%%%%%%%%%%%%%%%%%%%
w = 2*pi*f/fref;
H = (A+B*exp(-1i*w))./(1-C*exp(-1i*w));
Hint = (Aint+Bint*exp(-1i*w))./(1-Cint*exp(-1i*w));
Hextra =
   (Aextra+Bextra*exp(-1i*w))./(1-Cextra*exp(-1i*w));
%%%% Feedforward and feedback transfer functions for
    the phase noise
%%%% of the reference oscillator
HffRef =  kv*kp * klf * H .* Hint .* Hextra; %
    Hfeedforward
HfbRef = 1/N; % Hfeedback
HclRef =  HffRef./(1+HffRef.*HfbRef);  % Closed Loop TF
     for the phase noise of Ref
```

```
%%%% Feedforward and feedback transfer functions for
   noise at the DCOinput
HffDCO =  Hint*kv; % Hfeedforward
HfbDCO  = (1/N) * kp * klf * H .* Hextra; % Hfeedback
HclDCO =  HffDCO./(1+HffDCO.*HfbDCO);  % Closed Loop TF
     for the noise at the DCO input

VarTDCquant = (pi^2)*(fref).^2*(resTDC^2)/12;
PSDinputDCO_TDCquantComponent =
   (abs(HclRef./(Hint*kv))).^2 * VarTDCquant;
accumulationGain = (1/(Ts*2*pi)); % The sdomain
   transfer function of an accumulator that is clocked
   by
% a signal with period Ts is TFintegrator = (
   (1/Ts)/(2*pi) )/f
PSDphasePredicted_TDCquant = (1./f).^2.*
    (Ts*2*pi*kv)^2 *accumulationGain^2*
    Ts.*PSDinputDCO_TDCquantComponent;
PSDinputDCO_DCOcomponent = (abs(HclDCO./(Hint*kv))).^2
   * VarAdditiveNoiseDCOinput;
PSDphasePredicted_DCOinput = (1./f).^2.*
   (Ts*2*pi*kv)^2 *accumulationGain^2*
   Ts.*(PSDinputDCO_DCOcomponent);
VarDCOquant = (kdcoequivalent/kv).^2*(resDCO^2)/12;
PSDinputDCO_DCOquant = (abs(HclDCO./(Hint*kv))).^2 *
   VarDCOquant;
PSDphasePredicted_DCOquant = (1./f).^2.*
   (Ts*2*pi*kv)^2 *accumulationGain^2*
   Ts.*(PSDinputDCO_DCOquant);
%%%%% Plot Phase Noise %%%%%
figure
h=semilogx( f, 10*log10(PSDphase), 'b', f, 10*log10(
   PSDphasePredicted_TDCquant +
   PSDphasePredicted_DCOinput +
   PSDphasePredicted_DCOquant ), 'r');
grid on
set(gca, 'fontsize', 15, 'fontweight', 'bold')
xlabel('Frequency [Hz]')
ylabel('Phase Noise [dBc/Hz]')
grid on
set(h(1), 'LineWidth', 2)
set(h(2), 'LineWidth', 2)
axis([1e4 0.5e7 -150 -104])
legend('Matlab Model','Analytical Predictions')
```

Listing A.1 Matlab script for modeling a TDC-based ADPLL which includes a zeroth-order noise-shaping TDC and a DCO, each with resolution equal to the machine precision

```
%%%%%%%%%%%%%%%%%%%%%%%%%%%%%%%%%%%%%
%%%    PLL MODEL WHICH INCLUDES    %%%%
%%%    A TDC WITH QUANTIZATION     %%%%
%%%          AND DITHER            %%%%
```

```
%%%%%%%%%%%%%%%%%%%%%%%%%%%%%%%%%%
clc
close all
clear all
%%%%%%%%%%%%%%%%%%%%%%%%%%%%%%%%%%
%%% Parameters obtained from   %%%%%
%%%%%  PLL Assistant Designer  %%%%%
%%%%%%%%%%%%%%%%%%%%%%%%%%%%%%%%%%
% fz/fo = 0.22
% order 2
% shape Butter
% Ref freq 10e6
% out freq = 1e9
% fz = 2.2e5;
% wz = 2*pi*fz;
% fp = 1.734e5;
% wp = 2*pi*fp;
% N = 100;
% Kccpsim = 7.272e10;
%%%%%%%%%%%%%%%%%%%%%%%%%%%%%%%%%%%
Nsamples = 2^22;
fz = 2.2e4;
wz = 2*pi*fz;
fp = 1.734e5;
wp = 2*pi*fp;
fref = 10e6;
N = 100;
Kccpsim = 7.272e10;
KccpsimAdapted = Kccpsim/wz;
kv = 10e6;
kp = 1;
klf = N*KccpsimAdapted/(kv*kp);
%%%%%%%%%%%%%%%%%%%%%%%%
%%%  Inizialization   %%%
%%%%%%%%%%%%%%%%%%%%%%%%
freq_freerunning = fref*N;
output_phase_dco = zeros(Nsamples,1);
phase_input = zeros(Nsamples,1);
phase_err = zeros(Nsamples,1);
freq_dco = zeros(Nsamples,1);
out = zeros(Nsamples,1);
outExtra = zeros(Nsamples,1);
outExtraGain = zeros(Nsamples,1);
input_dco = zeros(Nsamples,1);
output_phase_dco(1) = 1;
freq_dco(1) = freq_freerunning;
eq = zeros(1,Nsamples);
tin = zeros(1,Nsamples);
outTDC = zeros(1,Nsamples);
outTDCwithGain = zeros(1,Nsamples);
eqDeltaC = zeros(1,Nsamples);
Finput = fref * ones(Nsamples,1); % Input Frequency
```

```
time_array =(1/fref) * (0:1: Nsamples-1); % Time Array
fs = fref; % Sampling Frequency of the Digital Filter
ditherDCO = zeros(1,length(time_array));
%%%%%%%%%%%%%%%%%%%%%%%%%%%%%%%%%%%%%%%%%%%%%%%%%%%%%%%%%%%%%%%
%%%%%%%  FILTER DESIGN: ZERO plus INTEGRATOR  %%%%%%%%%
%%%%%%%%%%%%%%%%%%%%%%%%%%%%%%%%%%%%%%%%%%%%%%%%%%%%%%%%%%%%%%%
syms z z1 outz inz p1;
s = 2 * fs * (z-1)/(z+1);
Hz = (s + z1) / (s); Hz = simple(Hz);
[N,D] = numden(Hz);
Cnum = coeffs(N, z);
Cden = coeffs(D, z);
Cnum1 = Cnum(1);
Cnum2 = Cnum(2);
Cden1 = Cden(1);
Cden2 = Cden(2);
fsval = 2*pi*fs; % sampling frequency in the discrete
    time domain
z1val =  wz;
Cnum1val = double(subs(Cnum1, {z1, fs},{z1val,
   fsval}));
Cnum2val = double(subs(Cnum2, {z1, fs},{z1val,
   fsval}));
Cden1val = double(subs(Cden1, {z1, fs},{z1val,
   fsval}));
Cden2val = double(subs(Cden2, {z1, fs},{z1val,
   fsval}));
A = (Cnum2val/ Cden2val);
B = (Cnum1val/ Cden2val);
C =  - (Cden1val/ Cden2val);
%%%%%%%%%%%%%%%%%%%%%%%%%%%%%%%%%%%%%%%%%%%%%%
%%%%%%%  FILTER DESIGN: EXTRA POLE  %%%%%%%%%
%%%%%%%%%%%%%%%%%%%%%%%%%%%%%%%%%%%%%%%%%%%%%%
s = 2 * fs * (z-1)/(z+1);
Hz = p1/(s+p1);
[N,D] = numden(Hz);
Cnum = coeffs(N, z);
Cden = coeffs(D, z);
Cnum1 = Cnum(1);
Cnum2 = Cnum(2);
Cden1 = Cden(1);
Cden2 = Cden(2);
p1val = wp;
fsval = 2*pi*fs; % sampling frequency in the discrete
    time domain
Cnum1valextra = double(subs(Cnum1, {p1, fs},{p1val,
   fsval}));
Cnum2valextra = double(subs(Cnum2, {p1, fs},{p1val,
   fsval}));
Cden1valextra = double(subs(Cden1, {p1, fs},{p1val,
   fsval}));
Cden2valextra = double(subs(Cden2, {p1, fs},{p1val,
```

```
   fsval})));
Aextra = (Cnum2valextra/ Cden2valextra);
Bextra = (Cnum1valextra/ Cden2valextra);
Cextra =  - (Cden1valextra/ Cden2valextra);
%%%%%%%%%%%%%%%%%%%%%%%%%%%%%%%%%%%%%%%%%%%%%%%
%%%%%%%  FILTER DESIGN: INTEGRATOR  %%%%%%%%%
%%%%%%%%%%%%%%%%%%%%%%%%%%%%%%%%%%%%%%%%%%%%%%%
s = 2 * fs * (z-1)/(z+1);
Hz = 1/s; [N,D] = numden(Hz);
Cnum = coeffs(N, z);
Cden = coeffs(D, z);
Cnum1 = Cnum(1);
Cnum2 = Cnum(2);
Cden1 = Cden(1);
Cden2 = Cden(2);
fsval = 2*pi*fs; % sampling frequency in the discrete
    time domain
Cnum1valint = double(subs(Cnum1,  fs, fsval));
Cnum2valint = double(subs(Cnum2,  fs, fsval));
Cden1valint = double(subs(Cden1,  fs, fsval));
Cden2valint = double(subs(Cden2,  fs, fsval));
Aint = (Cnum2valint/ Cden2valint);
Bint = (Cnum1valint/ Cden2valint);
Cint =  - (Cden1valint/ Cden2valint);
%%%%%%%%%%%%%%%%%%%%%%%%%%%%%%%%%%%%%%%%%%%%%%%%%%%%%%%%%%
%%%%%% Noise to be added to the input of the DCO %%%%%%
%%%%%%%%%%%%%%%%%%%%%%%%%%%%%%%%%%%%%%%%%%%%%%%%%%%%%%%%%%
foff = 1e6; % [Hz]
noise_at_foff = -130; % [dBc/Hz]
noise_var = (foff^2)/(kv^2)*10^(noise_at_foff/10);
Ts = 1/fref;
AdditiveNoiseDCOinput =
   sqrt(noise_var/Ts)*randn(1,length(time_array));
VarAdditiveNoiseDCOinput = var(AdditiveNoiseDCOinput);
%%%%%%%%%%%%%%%%%%%%%%%%%%%%%%%%%%%%%%%%%%%%%%%%%%%%%%%%%%
%%%%%%      Dither added to the TDC input        %%%%%%
%%%%%%%%%%%%%%%%%%%%%%%%%%%%%%%%%%%%%%%%%%%%%%%%%%%%%%%%%%
swichTDCdtr = 1; % set swichTDCdtr = 0 in order to
   switch off the TDC dither
swichDCOdtr = 1; % set swichTDCdtr = 0 in order to
   switch off the DCO dither
dither =
   swichTDCdtr*(1/3)*randn(1,length(time_array)); % TDC
        dither

for n = 2:1:length(time_array)
%%%%%%%%%%%%%%%%%%%%%%%%%%%%%%%%%%%%%%%%
%%%   Freq to Phase conversion      %%%
%%%        of the input Fref        %%%
%%%%%%%%%%%%%%%%%%%%%%%%%%%%%%%%%%%%%%%%
phase_input(n) =  (Cnum2valint/ Cden2valint) *
   Finput(n)  + (Cnum1valint/ Cden2valint) *
```

```
   Finput(n-1) - (Cden1valint/ Cden2valint) *
   phase_input(n-1) ;
if n == 2
   N = 100;
end
%%%%%%%%%%%%%%
%%%   TDC   %%%
%%%%%%%%%%%%%%
resTDC = 10e-12;
phase_err(n) = (phase_input(n) - output_phase_dco(n
   -1));
tin(n) = (1/fref)* phase_err(n);
tin(n) = tin(n) + resTDC*dither(n);
outTDC(n) = resTDC*floor(tin(n)/resTDC);
eq(n) = resTDC*(tin(n)/resTDC - floor(tin(n)/resTDC));
outTDCwithGain(n) = kp* outTDC(n)*fref;
%%%%%%%%%%%%%%%%%%%%%%%%%%%%%%%%%%%%%%
%%%  FILTER: zero plus integrator  %%%
%%%%%%%%%%%%%%%%%%%%%%%%%%%%%%%%%%%%%%
out(n)  =   Cnum2val/Cden2val * outTDCwithGain(n) +
   Cnum1val/ Cden2val * outTDCwithGain(n-1) -
   Cden1val/ Cden2val   * out(n-1);
%%%%%%%%%%%%%%%%%%%%%%%%%%%%%%%%%%%%%%
%%%      FILTER: Extra Pole        %%%
%%%%%%%%%%%%%%%%%%%%%%%%%%%%%%%%%%%%%%
outExtra(n) =   (Cnum2valextra/ Cden2valextra) *
   out(n)  + (Cnum1valextra/ Cden2valextra) * out(n-1)
   - (Cden1valextra/ Cden2valextra) * outExtra(n-1) ;
%%%%%%%%%%%%%%%%%%%%%%%%%%%%%%%%%%%%%%%%%%%%%%%
%%%%         Digital Filter Gain          %%%%%
%%%%%%%%%%%%%%%%%%%%%%%%%%%%%%%%%%%%%%%%%%%%%%%
outExtraGain(n) = klf * outExtra(n);
input_dco(n) = outExtraGain(n); % +
   AdditiveNoiseDCOinput(n);
%%%%%%%%%%%%%%%%%%%%%%%%%%%%%%%%%%%%%%%%%%%%%%%
%%%%        DCO with quantization         %%%%%
%%%%%%%%%%%%%%%%%%%%%%%%%%%%%%%%%%%%%%%%%%%%%%%
c = 10e-12;
l = 100e-9;
kdcoequivalent = -1/(2*sqrt((l*c)^3));
deltaC = kv*(1/kdcoequivalent) * input_dco(n);
resDCO = 1e-20;  % [Farads] 1e-20 F is negligible
ditherDCO(n) = swichDCOdtr* floor(8*rand(1,1)-4);
deltaCq = resDCO * (floor(deltaC/resDCO) +
   ditherDCO(n));
eqDeltaC(n) = resDCO*(deltaC/resDCO -
   floor(deltaC/resDCO));
freq_dco(n) = 1/sqrt(l*c) +  kdcoequivalent*deltaCq  +
   kv*AdditiveNoiseDCOinput(n);
output_phase_dco(n) =  (Cnum2valint/ Cden2valint) *
   freq_dco(n)/N  + (Cnum1valint/ Cden2valint) *
   freq_dco(n-1)/N - (Cden1valint/ Cden2valint) *
```

```
   output_phase_dco(n-1);
end

%%%%% Selection of the steady state samples    %%%%%
SteadyStateOffset = 10e5;
input_dco_steady = input_dco(SteadyStateOffset:end);
outExtraGain_steady =
   outExtraGain(SteadyStateOffset:end);
freq_dco = freq_dco(SteadyStateOffset:end);
%%%%%    Calculate the DCO's Phase Noise      %%%%%
num_segments = 16;
vin =  freq_dco/kv;   %; input_dco;
phase = filter(Ts*2*pi*kv,[1 -1],vin-mean(vin));
window_length = floor(length(phase)/num_segments);
[PSDphase,f] =
   pwelch(phase,window_length,[],[],1/Ts,'twosided');
%%%%%%%%%%%%%%%%%%%%%%%%%%%%%%%%%%%%%%%%%%%%%%%%%%%
%%%%%         Calculate the PSD of            %%%%%
%%%%%        the quantization error           %%%%%
%%%%%%%%%%%%%%%%%%%%%%%%%%%%%%%%%%%%%%%%%%%%%%%%%%%
window_length = floor(length(eq)/num_segments);
[PSDeq,feq] =
   pwelch(eq,window_length,[],[],1/Ts,'twosided');
%%%%%%%%%%%%%%%%%%%%%%%%%%%%%%%%%%%%%%%%%%%%%%%%%%%
%%%%%         Analytical Preditctions         %%%%%
%%%%%%%%%%%%%%%%%%%%%%%%%%%%%%%%%%%%%%%%%%%%%%%%%%%
w = 2*pi*f/fref;
H = (A+B*exp(-1i*w))./(1-C*exp(-1i*w));
Hint = (Aint+Bint*exp(-1i*w))./(1-Cint*exp(-1i*w));
Hextra =
   (Aextra+Bextra*exp(-1i*w))./(1-Cextra*exp(-1i*w));
%%%% Feedforward and feedback transfer functions for
    the phase noise
%%%% of the reference oscillator
HffRef =  kv*kp * klf * H .* Hint .* Hextra; %
    Hfeedforward
HfbRef = 1/N; % Hfeedback
HclRef =  HffRef./(1+HffRef.*HfbRef);  % Closed Loop TF
     for the phase noise of Ref
%%%% Feedforward and feedback transfer functions for
    noise at the DCOinput
HffDCO =  Hint*kv; % Hfeedforward
HfbDCO  = (1/N) * kp * klf * H .* Hextra; % Hfeedback
HclDCO =  HffDCO./(1+HffDCO.*HfbDCO);  % Closed Loop TF
     for the noise at the DCO input

VarTDCadditiveNoise =
   (fref).^2*((swichTDCdtr*resTDC)^2)/12;
VarTDCquant = (fref).^2*(resTDC^2)/12;
PSDinputDCO_TDCquantComponent =
   (abs(HclRef)./abs(Hint*kv)).^2 * ( VarTDCquant) ;
PSDinputDCO_TDCAdditive =
```

```
   (abs(HclRef)./abs(Hint*kv)).^2 * (
   VarTDCadditiveNoise);
accumulationGain = (1/(Ts*2*pi)); % The sdomain
   transfer function of an accumulator that is clocked
   by
% a signal with period Ts is TFintegrator = (
   (1/Ts)/(2*pi) )/f
PSDphasePredicted_TDCquant = (1./f).^2.*
   (Ts*2*pi*kv)^2 *accumulationGain^2*
   Ts.*PSDinputDCO_TDCquantComponent;
PSDphasePredicted_TDCAdditive = (1./f).^2.*
   (Ts*2*pi*kv)^2 *accumulationGain^2*
   Ts.*PSDinputDCO_TDCAdditive;
PSDinputDCO_DCOcomponent = (abs(HclDCO./(Hint*kv))).^2
   * VarAdditiveNoiseDCOinput;
PSDphasePredicted_DCOinput = (1./f).^2.*
   (Ts*2*pi*kv)^2 *accumulationGain^2*
   Ts.*(PSDinputDCO_DCOcomponent);
VarDCOquant = (kdcoequivalent/kv).^2*(resDCO^2)/12;
PSDinputDCO_DCOquant = (abs(HclDCO./(Hint*kv))).^2 *
   VarDCOquant;
PSDphasePredicted_DCOquant = (1./f).^2.*
   (Ts*2*pi*kv)^2 *accumulationGain^2*
   Ts.*(PSDinputDCO_DCOquant);
PSDeqPredicted =
    (fref).^2*Ts*ones(1,length(PSDeq))*((resTDC)^2)/12;
%%%%% Plot PSD Quantization error %%%%%
figure h = semilogx( feq, 10*log10((fref).^2*PSDeq),
   'b',
          feq, 10*log10(PSDeqPredicted), 'r');
set(gca, 'fontsize', 15, 'fontweight', 'bold')
xlabel('Frequency [Hz]')
ylabel('PSD [dB/Hz]')
grid on
set(h(1), 'LineWidth', 6)
set(h(2), 'LineWidth', 4)
axis([1e3 0.5e7 -190 -130])
legend('Matlab Model','White Noise Approximation')
%%%%% Plot Phase Noise %%%%%
figure
h=semilogx( f, 10*log10(PSDphase), 'b', f, 10*log10(
   PSDphasePredicted_TDCAdditive +
   PSDphasePredicted_TDCquant +
   PSDphasePredicted_DCOinput +
   PSDphasePredicted_DCOquant ), 'r', ...
   f, 10*log10(PSDphasePredicted_DCOinput),  'g--', ...
   f, 10*log10(PSDphasePredicted_TDCquant +
      PSDphasePredicted_TDCAdditive), 'g');
grid on
set(gca, 'fontsize', 15, 'fontweight', 'bold')
xlabel('Frequency [Hz]')
ylabel('Phase Noise [dBc/Hz]')
```

```
grid on
set(h(1), 'LineWidth', 9)
set(h(2), 'LineWidth', 8)
set(h(3), 'LineWidth', 2)
set(h(4), 'LineWidth', 3)
axis([1e3 0.5e7 -150 -74])
legend('Matlab Model','Analytical Predictions', 'DCO
   noise', 'TDC noise')
```

Listing A.2 Matlab script for modeling a TDC-based ADPLL which includes a zeroth-order noise-shaping TDC with finite resolution and a DCO with resolution equal to the machine precision. Dither is added to the TDC input

```
%%%%%%%%%%%%%%%%%%%%%%%%%%%%%%%%%%%%
%%%     PLL MODEL WHICH INCLUDES   %%%%
%%% A FIRST-ORDER NOISE-SHAPING %%%%
%%%        TDC WITHOUT DITHER      %%%%
%%%%%%%%%%%%%%%%%%%%%%%%%%%%%%%%%%%%
clc
close all
clear all
%%%%%%%%%%%%%%%%%%%%%%%%%%%%%%%%%%%%
%%% Parameters obtained from    %%%%%
%%%%%   PLL Assistant Designer   %%%%%
%%%%%%%%%%%%%%%%%%%%%%%%%%%%%%%%%%%%
% fz/fo = 0.22
% order 2
% shape Butter
% Ref freq 10e6
% out freq = 1e9
% fz = 2.2e5;
% wz = 2*pi*fz;
% fp = 1.734e5;
% wp = 2*pi*fp;
% N = 100;
% Kccpsim = 7.272e10;
%%%%%%%%%%%%%%%%%%%%%%%%%%%%%%%%%%%%%
Nsamples = 2^22;
fz = 2.2e4;
wz = 2*pi*fz;
fp = 1.734e5;
wp = 2*pi*fp;
fref = 10e6;
N = 100;
Kccpsim = 7.272e10;
KccpsimAdapted = Kccpsim/wz;
kv = 10e6;
kp = 1;
klf = N*KccpsimAdapted/(kv*kp);
%%%%%%%%%%%%%%%%%%%%%%%%%
%%%  Inizialization    %%%
%%%%%%%%%%%%%%%%%%%%%%%%%
```

```
freq_freerunning = fref*N; % freerunning frequency of
    the controlled oscillator
output_phase_dco = zeros(Nsamples,1);
phase_input = zeros(Nsamples,1);
phase_err = zeros(Nsamples,1);
freq_dco = zeros(Nsamples,1);
out = zeros(Nsamples,1);
outExtra = zeros(Nsamples,1);
outExtraGain = zeros(Nsamples,1);
input_dco = zeros(Nsamples,1);
output_phase_dco(1) = 1;
freq_dco(1) = freq_freerunning;
Finput = fref * ones(Nsamples,1); % Input Frequency
time_array =(1/fref) * (0:1: Nsamples-1); % Time Array
eq = zeros(1,Nsamples);
tin = zeros(1,Nsamples);
outTDC = zeros(1,Nsamples);
outTDCwithGain = zeros(1,Nsamples);
eqDeltaC = zeros(1,Nsamples);
fs = fref; % Sampling Frequency of the Digital Filter
%%%%%%%%%%%%%%%%%%%%%%%%%%%%%%%%%%%%%%%%%%%%%%%%%%%%%%%
%%%%%%  FILTER DESIGN: ZERO plus INTEGRATOR  %%%%%%%%
%%%%%%%%%%%%%%%%%%%%%%%%%%%%%%%%%%%%%%%%%%%%%%%%%%%%%%%
syms z z1 outz inz p1;
s = 2 * fs * (z-1)/(z+1);
Hz = (s + z1) / (s);
Hz = simple(Hz);
[N,D] = numden(Hz);
Cnum = coeffs(N, z);
Cden = coeffs(D, z);
Cnum1 = Cnum(1);
Cnum2 = Cnum(2);
Cden1 = Cden(1);
Cden2 = Cden(2);
fsval = 2*pi*fs; % sampling frequency in the discrete
    time domain
z1val = wz;
Cnum1val = double(subs(Cnum1, {z1, fs},{z1val,
    fsval}));
Cnum2val = double(subs(Cnum2, {z1, fs},{z1val,
    fsval}));
Cden1val = double(subs(Cden1, {z1, fs},{z1val,
    fsval}));
Cden2val = double(subs(Cden2, {z1, fs},{z1val,
    fsval}));
A = (Cnum2val/ Cden2val);
B = (Cnum1val/ Cden2val);
C = - (Cden1val/ Cden2val);
%%%%%%%%%%%%%%%%%%%%%%%%%%%%%%%%%%%%%%%%%%%%%
%%%%%%  FILTER DESIGN: EXTRA POLE  %%%%%%%%
%%%%%%%%%%%%%%%%%%%%%%%%%%%%%%%%%%%%%%%%%%%%%
s = 2 * fs * (z-1)/(z+1);
```

```
Hz = p1/(s+p1);
[N,D] = numden(Hz);
Cnum = coeffs(N, z);
Cden = coeffs(D, z);
Cnum1 = Cnum(1);
Cnum2 = Cnum(2);
Cden1 = Cden(1);
Cden2 = Cden(2);
p1val = wp;
fsval = 2*pi*fs; % sampling frequency in the discrete
    time domain
Cnum1valextra = double(subs(Cnum1, {p1, fs},{p1val,
    fsval}));
Cnum2valextra = double(subs(Cnum2, {p1, fs},{p1val,
    fsval}));
Cden1valextra = double(subs(Cden1, {p1, fs},{p1val,
    fsval}));
Cden2valextra = double(subs(Cden2, {p1, fs},{p1val,
    fsval}));
Aextra = (Cnum2valextra/ Cden2valextra);
Bextra = (Cnum1valextra/ Cden2valextra);
Cextra =  - (Cden1valextra/ Cden2valextra);
%%%%%%%%%%%%%%%%%%%%%%%%%%%%%%%%%%%%%%%%
%%%%%%  FILTER DESIGN: INTEGRATOR  %%%%%%%%
%%%%%%%%%%%%%%%%%%%%%%%%%%%%%%%%%%%%%%%%
s = 2 * fs * (z-1)/(z+1); Hz = 1/s;
[N,D] = numden(Hz);
Cnum = coeffs(N, z);
Cden = coeffs(D, z);
Cnum1 = Cnum(1);
Cnum2 = Cnum(2);
Cden1 = Cden(1);
Cden2 = Cden(2);
fsval = 2*pi*fs; % sampling frequency in the discrete
    time domain
Cnum1valint = double(subs(Cnum1,  fs, fsval));
Cnum2valint = double(subs(Cnum2,  fs, fsval));
Cden1valint = double(subs(Cden1,  fs, fsval));
Cden2valint = double(subs(Cden2,  fs, fsval));
Aint = (Cnum2valint/ Cden2valint);
Bint = (Cnum1valint/ Cden2valint);
Cint =  - (Cden1valint/ Cden2valint);
%%%%%%%%%%%%%%%%%%%%%%%%%%%%%%%%%%%%%%%%%%%%%%%%%
%%%%% Noise to be added to the input of the DCO %%%%%
%%%%%%%%%%%%%%%%%%%%%%%%%%%%%%%%%%%%%%%%%%%%%%%%%
foff = 1e6; % [Hz]
noise_at_foff = -130; % [dBc/Hz]
% noise_at_foff = -100; % dBc
noise_var = (foff^2)/(kv^2)*10^(noise_at_foff/10);
Ts = 1/fref;
AdditiveNoiseDCOinput =
     sqrt(noise_var/Ts)*randn(1,length(time_array));
```

```
VarAdditiveNoiseDCOinput = var(AdditiveNoiseDCOinput);
%%%%%%%%%%%%%%%%%%%%%%%%%%%%%%%%%%%%%%%%%%%%%%%%%%%%%%%%%%%%%%%%%
%%%%%%      Dither added to the TDC input      %%%%%%
%%%%%%%%%%%%%%%%%%%%%%%%%%%%%%%%%%%%%%%%%%%%%%%%%%%%%%%%%%%%%%%%%%
swichTDCdtr = 0; % set swichTDCdtr = 0 in order to
   switch off the dither
dither =
   swichTDCdtr*(1/3)*randn(1,length(time_array)); %
      three sigma of this signal is equal to 1
for n = 2:1:length(time_array)
%%%%%%%%%%%%%%%%%%%%%%%%%%%%%%%%%%%%%%%%%%%%%%%%%%
%%%   Freq to Phase conversion     %%%
%%%        of the input Fref       %%%
%%%%%%%%%%%%%%%%%%%%%%%%%%%%%%%%%%%%%%%%%%%%%%%%%%
phase_input(n) =  (Cnum2valint/ Cden2valint) *
   Finput(n)  + (Cnum1valint/ Cden2valint) *
   Finput(n-1) - (Cden1valint/ Cden2valint) *
   phase_input(n-1) ;
if n == 2
   N = 100;
end
%%%%%%%%%%%%%%%%%%%%%%%%
%%%   TDC   %%%
%%%%%%%%%%%%%%%%%%%%%%%%
resTDC = 10e-12;
phase_err(n) = (phase_input(n) - output_phase_dco(n
   -1));
tin(n) = (1/fref)* phase_err(n);
tin(n) = tin(n) + resTDC*dither(n);
%%% Sigma-delta Modulator %%%
InputSigmaDelta = tin(n)/resTDC;
outTDC(n) = resTDC * floor( InputSigmaDelta + eq(n -1)
   );
eq(n) =   ((InputSigmaDelta + eq(n-1)) - floor(
    InputSigmaDelta + eq(n -1) ));
outTDCwithGain(n) = kp* outTDC(n)*fref;
%%%%%%%%%%%%%%%%%%%%%%%%%%%%%%%%%%%%%%%%%%%%%%%%%%
%%%  FILTER: zero plus integrator  %%%
%%%%%%%%%%%%%%%%%%%%%%%%%%%%%%%%%%%%%%%%%%%%%%%%%%
out(n)  =   Cnum2val/Cden2val * outTDCwithGain(n) +
    Cnum1val/ Cden2val * outTDCwithGain(n-1) -
    Cden1val/ Cden2val   * out(n-1);
%%%%%%%%%%%%%%%%%%%%%%%%%%%%%%%%%%%%%%%%%%%%%%%%%%
%%%      FILTER: Extra Pole          %%%
%%%%%%%%%%%%%%%%%%%%%%%%%%%%%%%%%%%%%%%%%%%%%%%%%%
outExtra(n) =   (Cnum2valextra/ Cden2valextra) *
    out(n)  + (Cnum1valextra/ Cden2valextra) * out(n-1)
    - (Cden1valextra/ Cden2valextra) * outExtra(n-1) ;
%%%%%%%%%%%%%%%%%%%%%%%%%%%%%%%%%%%%%%%%%%%%%%%%%%%%%%%%%%%%%%%%
%%%%         Digital Filter Gain          %%%%%
%%%%%%%%%%%%%%%%%%%%%%%%%%%%%%%%%%%%%%%%%%%%%%%%%%%%%%%%%%%%%%%%
outExtraGain(n) = klf * outExtra(n);
```

```
input_dco(n) = outExtraGain(n); % +
   AdditiveNoiseDCOinput(n);
%%%%%%%%%%%%%%%%%%%%%%%%%%%%%%%%%%%%%%%%%%%%%%%%
%%%%          DCO with quantization        %%%%%
%%%%%%%%%%%%%%%%%%%%%%%%%%%%%%%%%%%%%%%%%%%%%%%%
c = 10e-12;
l = 100e-9;
kdcoequivalent = -1/(2*sqrt((l*c)^3));
deltaC = kv*(1/kdcoequivalent) * input_dco(n);
resDCO = 1e-20; % [Farads] 1e-20 F is negligible
deltaCq = resDCO * floor(deltaC/resDCO);
eqDeltaC(n) = resDCO*(deltaC/resDCO -
   floor(deltaC/resDCO));
freq_dco(n) = 1/sqrt(l*c) +  kdcoequivalent*deltaCq  +
   kv*AdditiveNoiseDCOinput(n);
output_phase_dco(n) =  (Cnum2valint/ Cden2valint) *
   freq_dco(n)/N  + (Cnum1valint/ Cden2valint) *
   freq_dco(n-1)/N - (Cden1valint/ Cden2valint) *
   output_phase_dco(n-1);
end
%%%%%%%%%%%%%%%%%%%%%%%%%%%%%%%%%%%%%%%%%%%%%%%%%%
%%%%% Selection of the steady state samples  %%%%%
%%%%%%%%%%%%%%%%%%%%%%%%%%%%%%%%%%%%%%%%%%%%%%%%%%
SteadyStateOffset = 10e5;
input_dco_steady = input_dco(SteadyStateOffset:end);
outExtraGain_steady =
   outExtraGain(SteadyStateOffset:end);
freq_dco = freq_dco(SteadyStateOffset:end);
%%%%%%%%%%%%%%%%%%%%%%%%%%%%%%%%%%%%%%%%%%%%%%%%%%
%%%%%    Calculate the DCO's Phase Noise     %%%%%
%%%%%%%%%%%%%%%%%%%%%%%%%%%%%%%%%%%%%%%%%%%%%%%%%%
num_segments = 32;
vin =  freq_dco/kv;    %; input_dco;
phase = filter(Ts*2*pi*kv,[1 -1],vin-mean(vin));
window_length = floor(length(phase)/num_segments);
[PSDphase,f] =
   pwelch(phase,window_length,[],[],1/Ts,'twosided');
%%%%%%%%%%%%%%%%%%%%%%%%%%%%%%%%%%%%%%%%%%%%%%%%%%
%%%%%          Calculate the PSD of         %%%%%
%%%%%          the quantization error       %%%%%
%%%%%%%%%%%%%%%%%%%%%%%%%%%%%%%%%%%%%%%%%%%%%%%%%%
window_length = floor(length(eq)/num_segments);
[PSDeq,feq] = pwelch( resTDC*eq,
   window_length,[],[],1/Ts,'twosided');
%%%%%%%%%%%%%%%%%%%%%%%%%%%%%%%%%%%%%%%%%%%%%%%%%%
%%%%%          Analytical Preditctions      %%%%%
%%%%%%%%%%%%%%%%%%%%%%%%%%%%%%%%%%%%%%%%%%%%%%%%%%
w = 2*pi*f/fref;
H = (A+B*exp(-1i*w))./(1-C*exp(-1i*w));
Hint = (Aint+Bint*exp(-1i*w))./(1-Cint*exp(-1i*w));
Hextra =
   (Aextra+Bextra*exp(-1i*w))./(1-Cextra*exp(-1i*w));
```

```
%%%% Feedforward and feedback transfer functions for
    the phase noise
%%%% of the reference oscillator
HffRef =  kv*kp * klf * H .* Hint .* Hextra; %
    Hfeedforward
HfbRef = 1/N; % Hfeedback
HclRef =  HffRef./(1+HffRef.*HfbRef);  % Closed Loop TF
     for the phase noise of Ref
%%%% Feedforward and feedback transfer functions for
    noise at the DCOinput
HffDCO =  Hint*kv; % Hfeedforward
HfbDCO  = (1/N) * kp * klf * H .* Hextra; % Hfeedback
HclDCO =  HffDCO./(1+HffDCO.*HfbDCO);  % Closed Loop TF
     for the noise at the DCO input

VarTDCadditiveNoise =
     (fref).^2*((swichTDCdtr*resTDC)^2)/12;
VarTDCquant = (fref).^2*(resTDC^2)/12;
PSDinputDCO_TDCquantComponent =
    (abs(HclRef)./abs(Hint*kv)).^2 * ( VarTDCquant)
    .*(2-2*cos(w)) ;
PSDinputDCO_TDCAdditive =
     (abs(HclRef)./abs(Hint*kv)).^2 * (
     VarTDCadditiveNoise);
accumulationGain = (1/(Ts*2*pi)); % The sdomain
    transfer function of an accumulator that is clocked
    by
% a signal with period Ts is TFintegrator = (
    (1/Ts)/(2*pi) )/f
PSDphasePredicted_TDCquant = (1./f).^2.*
    (Ts*2*pi*kv)^2 *accumulationGain^2*
    Ts.*PSDinputDCO_TDCquantComponent;
PSDphasePredicted_TDCAdditive = (1./f).^2.*
    (Ts*2*pi*kv)^2 *accumulationGain^2*
    Ts.*PSDinputDCO_TDCAdditive;
PSDinputDCO_DCOcomponent = (abs(HclDCO./(Hint*kv))).^2
    * VarAdditiveNoiseDCOinput;
PSDphasePredicted_DCOinput = (1./f).^2.*
     (Ts*2*pi*kv)^2 *accumulationGain^2*
     Ts.*(PSDinputDCO_DCOcomponent);
VarDCOquant = (kdcoequivalent/kv).^2*(resDCO^2)/12;
PSDinputDCO_DCOquant = (abs(HclDCO./(Hint*kv))).^2 *
     VarDCOquant;
PSDphasePredicted_DCOquant = (1./f).^2.*
    (Ts*2*pi*kv)^2 *accumulationGain^2*
    Ts.*(PSDinputDCO_DCOquant);
PSDeqPredicted =
     (fref).^2*Ts*ones(1,length(PSDeq))*((resTDC)^2)/12;
%%%%% Plot PSD Quantization error %%%%%
figure
h = semilogx( feq, 10*log10((fref).^2*PSDeq), 'b',
    feq, 10*log10(PSDeqPredicted), 'r');
```

```
set(gca, 'fontsize', 15, 'fontweight', 'bold')
xlabel('Frequency [Hz]')
ylabel('PSD [dB/Hz]')
grid on
set(h(1), 'LineWidth', 6)
set(h(2), 'LineWidth', 4)
axis([1e3 0.5e7 -190 -130])
legend('Matlab Model','White Noise Approximation')
%%%%% Plot Phase Noise %%%%%
figure
h=semilogx( f, 10*log10(PSDphase), 'b', f,
   10*log10(PSDphasePredicted_TDCAdditive +
   PSDphasePredicted_TDCquant +
   PSDphasePredicted_DCOinput +
   PSDphasePredicted_DCOquant), 'r', ...
   f, 10*log10(PSDphasePredicted_DCOinput),  'g--', ...
   f, 10*log10(PSDphasePredicted_TDCquant +
      PSDphasePredicted_TDCAdditive), 'g');
grid on
set(gca, 'fontsize', 15, 'fontweight', 'bold')
xlabel('Frequency [Hz]')
ylabel('Phase Noise [dBc/Hz]')
grid on
set(h(1), 'LineWidth', 9)
set(h(2), 'LineWidth', 8)
set(h(3), 'LineWidth', 2)
set(h(4), 'LineWidth', 3)
axis([1e3 0.5e7 -150 -104])
legend('Matlab Model','Analytical Predictions', 'DCO
   noise', 'TDC noise')
```

Listing A.3 Matlab script for modeling a TDC-based ADPLL which includes a first-order noise-shaping TDC with finite resolution and a DCO with resolution equal to the machine precision. Dither is *not* added to the TDC input

A.2 Matlab Scripts for Phase Noise Analysis

```
clc
close all
clear all
tstep = 1;
nsamples = 2^23;
time = 0:tstep:nsamples-1;
period = 2^9;
ThreeSigma = period/20;
Sigma = ThreeSigma/3;
varjitter = Sigma^2;
Nperiods = length(time)/period +10;
```

```
jitterWhite = Sigma*randn(1,nsamples);
%%%%%%%%%%%%%%%%%%%%%%%%%%%%%%%%%%%%%%%%
%%%%%%%  FILTER DESIGN: Single POLE  %%%%%%%%%
%%%%%%%%%%%%%%%%%%%%%%%%%%%%%%%%%%%%%%%%
syms z z1 outz inz p1;
fs = tstep/period;
s = 2 * fs * (z-1)/(z+1);
Hz = p1/(s+p1);
[N,D] = numden(Hz);
Cnum = coeffs(N, z);
Cden = coeffs(D, z);
Cnum1 = Cnum(1);
Cnum2 = Cnum(2);
Cden1 = Cden(1);
Cden2 = Cden(2);
fsval = 2*pi*fs; % sampling frequency in the discrete
   time domain
p1val = fsval/100; % frequency of the pole
Cnum1valextra = double(subs(Cnum1, {p1, fs},{p1val,
   fsval}));
Cnum2valextra = double(subs(Cnum2, {p1, fs},{p1val,
   fsval}));
Cden1valextra = double(subs(Cden1, {p1, fs},{p1val,
   fsval}));
Cden2valextra = double(subs(Cden2, {p1, fs},{p1val,
   fsval}));
Aextra = (Cnum2valextra/ Cden2valextra);
Bextra = (Cnum1valextra/ Cden2valextra);
Cextra =  - (Cden1valextra/ Cden2valextra);
%%%%%%%%%%%%%%%%%%%%%%%%%%%%%%%%%
%%%     FILTER: Implementation     %%%
%%%%%%%%%%%%%%%%%%%%%%%%%%%%%%%%%
jitterFiltered = zeros(1,length(jitterWhite));
for n = 2:1:length(jitterWhite)
jitterFiltered(n) =   (Cnum2valextra/ Cden2valextra) *
jitterWhite(n)  + (Cnum1valextra/ Cden2valextra) *
jitterWhite(n-1) - (Cden1valextra/ Cden2valextra) *
jitterFiltered(n-1) ;
end
%%%%%%%%%%%%%%%%%%%%%%%%%%%%%%%%%
%%%   SECOND FILTER: Design and Implementation    %%%
%%%%%%%%%%%%%%%%%%%%%%%%%%%%%%%%%
jitterWhiteNew = Sigma*randn(1,nsamples);
syms z z1 outz inz p1;
fs = tstep/period;
s = 2 * fs * (z-1)/(z+1);
Hz = p1/(s+p1);
[N,D] = numden(Hz);
Cnum = coeffs(N, z);
Cden = coeffs(D, z);
Cnum1 = Cnum(1);
Cnum2 = Cnum(2);
```

```
Cden1 = Cden(1);
Cden2 = Cden(2);
fsval = 2*pi*fs; % sampling frequency in the discrete
    time domain
p1val = fsval/10000; % frequency of the pole (Hz)
Cnum1valextra = double(subs(Cnum1, {p1, fs},{p1val,
   fsval}));
Cnum2valextra = double(subs(Cnum2, {p1, fs},{p1val,
   fsval}));
Cden1valextra = double(subs(Cden1, {p1, fs},{p1val,
   fsval}));
Cden2valextra = double(subs(Cden2, {p1, fs},{p1val,
   fsval}));
Aextra = (Cnum2valextra/ Cden2valextra);
Bextra = (Cnum1valextra/ Cden2valextra);
Cextra =  - (Cden1valextra/ Cden2valextra);
jitterFilteredNew = zeros(1,length(jitterWhite));
for n = 2:1:length(jitterWhite)
jitterFilteredNew(n) =  (Cnum2valextra/ Cden2valextra)
   * jitterWhiteNew(n)  + (Cnum1valextra/
   Cden2valextra) * jitterWhiteNew(n-1) -
   (Cden1valextra/ Cden2valextra) *
   jitterFilteredNew(n-1) ;
end
jitter = 0.01*( 0.01*cumsum(jitterFilteredNew) +
   0.01*jitterFiltered );
sinsignalNoisy = sin(2*pi*(1./period).*time+ jitter);
```

Listing A.4 Matlab script that produces a sinusoid with phase noise

```
hannWindow = hann(round(length(time)/16));
[PSDoutweltch ,wappo] = pwelch(sinsignalNoisy,
   hannWindow);
PSDoutweltch = pi*PSDoutweltch;
f = 0:0.5/(length(PSDoutweltch) -1):0.5;
hannWindow = hann(round(length(jitter)/16)); \
[PSDphaseComponent ,wappo] = pwelch(jitter,
   hannWindow);
PSDphaseComponent = pi*PSDphaseComponent;
fJitter = (0:0.5/(length(PSDphaseComponent) -1):0.5) ;
fJitter = fJitter + 1/period;
fMirror = 0:0.5/(length(PSDoutweltch) -1):1/period;
temp = PSDphaseComponent(1:1:length(fMirror));
PSDphaseComponentMirror = temp(end:-1:1);
figure
h = semilogx(f, 10*log10(PSDoutweltch), 'b', fJitter,
   10*log10(PSDphaseComponent/2), 'r', ...
fMirror, 10*log10(PSDphaseComponentMirror/2), 'r' );
set(h(1), 'LineWidth', 6);
set(h(2), 'LineWidth', 2);
set(h(3), 'LineWidth', 2);
set(gca, 'fontsize', 15, 'fontweight', 'bold')
```

```
xlabel('Normalized Frequency')
ylabel('PSD [dB/sample]')
legend('Amplitude','Phase Deviations')
axis([1e-5  0.5 -120  45])
grid on
```

Listing A.5 Matlab script that compares the PSD of the amplitude and phase deviations of a noisy signal

```
clc
close all
clear all
tstep = 1;
nsamples = 2^20;
time = 0:tstep:nsamples-1;
period = 2^9;
osc = sin(2*pi*(1/period)*time);
hannWindow = hann(round(length(time)/32));
[PSDoutweltch ,wappo] = pwelch(osc, hannWindow);
PSDoutweltch = pi*PSDoutweltch;
fnorm = 0:0.5/(length(PSDoutweltch) -1):0.5;
figure
h = semilogx(fnorm, 10*log10(PSDoutweltch), 'b',
    'LineWidth',4);
set(gca, 'fontsize', 15, 'fontweight', 'bold')
xlabel('Normalized Frequency')
ylabel('PSD [dB/sample]')
axis([1e-5  0.5 -120  45])
grid on
```

Listing A.6 Matlab script that shows the Power Spectral Density of a sinusoid without phase noise

```
%%%%%%%% Jitter extraction %%%%%%%%%%%
%%%%%% Processing the Periods %%%%%%%
idxCrossingsPositive = [];
idxCrossingsNegative = [];
for idx = 2:1:nsamples
%%%% rising edge detection %%%%
if (sinsignalNoisy(idx) > 0 ) && ( sinsignalNoisy(idx
   -1) < 0 )
idxCrossingsPositive = [idxCrossingsPositive idx];
idxCrossingsNegative = [idxCrossingsNegative (idx-1)];
end
end
grid on
%%%% Interpolation  %%%%%
xa = time(idxCrossingsNegative);
xb = time(idxCrossingsPositive);
ya = sinsignalNoisy(idxCrossingsNegative);
yb = sinsignalNoisy(idxCrossingsPositive);
te = xb - (xb - xa)/(yb - ya)*yb; % crossings instants
teideal = period*(1:1:length(te)); % ideal crossings
```

```
deltaT = (teideal - te);
deltaPhi = 2*pi*deltaT./(period);  % Calculate DeltaPhi
     from DeltaT
jitterExtracted = deltaPhi;
figure
h=plot(1:1:length(jitter), jitter, '-bo',
   (1:1:length(jitterExtracted))
   /length(jitterExtracted)*length(jitter),
   (jitterExtracted), 'r.'  );
set(h(1), 'LineWidth', 4);
set(h(2), 'MarkerSize', 5);
set(gca,  'fontsize', 15, 'fontweight', 'bold')
xlabel('Time Index')
ylabel('Phase Deviation')
legend('Actual','Extracted')
grid on
```

Listing A.7 Matlab script that compares the actual and the extracted phase variations of a noisy sinusoid

```
hannWindow = hann(round(length(jitterExtracted)/32));
[PSDjitterExtracted ,wappo] = pwelch(jitterExtracted
   , hannWindow);
PSDjitterExtracted = pi*PSDjitterExtracted;
Nperiods = nsamples/period;
SamplesPerPeriod = nsamples/Nperiods;
fjitterExtracted =
    (0:(0.5/SamplesPerPeriod)/(length(PSDjitterExtracted)
    -1):(0.5/SamplesPerPeriod)) ;
fjitterExtracted = fjitterExtracted + 1/period;
figure
h = semilogx(fJitter, 10*log10(PSDphaseComponent/2),
    'b', fjitterExtracted, 10*log10(SamplesPerPeriod
    *PSDjitterExtracted/2), 'g');
set(h(1), 'LineWidth', 6);
set(h(2), 'LineWidth', 2);
set(gca, 'fontsize', 15, 'fontweight', 'bold')
xlabel('Normalized Frequency')
ylabel('PSD [dB/sample]')
legend('Phase Deviation','Extracted Phase Deviation')
axis([(1/period)-3e-4  0.5 -120  45])
grid on
```

Listing A.8 Matlab script that compares the PSDs of actual and the extracted phase deviations of a noisy sinusoid

Index

F. Brandonisio and M. P. Kennedy, *Noise-Shaping All-Digital Phase-Locked Loops*,
Analog Circuits and Signal Processing, DOI: 10.1007/978-3-319-03659-5,

MIX
Papier aus verantwortungsvollen Quellen
Paper from responsible sources
FSC® C105338

If you have any concerns about our products,
you can contact us on
ProductSafety@springernature.com

In case Publisher is established outside the EU,
the EU authorized representative is:
Springer Nature Customer Service Center GmbH
Europaplatz 3, 69115 Heidelberg, Germany

Printed by Libri Plureos GmbH
in Hamburg, Germany